# Current Topics in Microbiology 249 and Immunology

Editors

R.W. Compans, Atlanta/Georgia
M. Cooper, Birmingham/Alabama · Y. Ito, Kyoto
H. Koprowski, Philadelphia/Pennsylvania · F. Melchers, Basel
M. Oldstone, La Jolla/California · S. Olsnes, Oslo
M. Potter, Bethesda/Maryland
P.K. Vogt, La Jolla/California · H. Wagner, Munich

**Springer**
Berlin
Heidelberg
New York
Barcelona
Hong Kong
London
Milan
Paris
Singapore
Tokyo

# DNA Methylation and Cancer

Edited by P.A. Jones and
P.K. Vogt

With 16 Figures and 11 Tables

 Springer

Dr. PETER A. JONES
Director
Norris Comprehensive Cancer Center
and Hospital
University of Southern California
Room 8315, Mail Stop #83
1441 Eastlake Ave.
Los Angeles, CA 90089-9181
USA
*E-mail*: jones_p@ccnt.hsc.usc.edu

Professor PETER K. VOGT, Ph.D.
Division of Oncovirology
Mail Drop & Room BCC239
Department of Molecular
and Experimental Medicine
The Scripps Research Institute
10550 N. Torrey Pines Road
La Jolla, CA 92037
USA
*E-mail*: pkvogt@scripps.edu

*Cover Illustration:* "Knudson's two-hit hypothesis revised. Two active alleles of a tumour-suppressor gene are indicated by the two green boxes shown at the top. The first step of gene inactivation is shown as a localized mutation on the left or by transcriptional repression by DNA methylation on the right. The second hit is shown by either LOH or transcriptional silencing. (Jones, P.A. and Laird, P.W., Nature Genetics 21:163–167, 1999)."

ISSN 0070-217X
ISBN-13: 978-3-642-64090-2 Springer-Verlag Berlin Heidelberg New York

ISBN-13: 978-3-642-64090-2     e-ISBN-13: 978-3-642-59696-4
DOI: 10.1007/978-3-642-59696-4

© Springer-Verlag Berlin Heidelberg 2000
Softcover reprint of the hardcover 1st edition 2000

Library of Congress Catalog Card Number 15-12910

The use of general descriptive names, registered names, trademarks, etc. in this publication does not imply, even in the absence of a specific statement, that such names are exempt from the relevant protective laws and regulations and therefore free for general use.

Product liability: The publishers cannot guarantee the accuracy of any information about dosage and application contained in this book. In every individual case the user must check such information by consulting other relevant literature.

Cover Design: *design & production GmbH*, Heidelberg
Typesetting: Scientific Publishing Services (P) Ltd, Madras
Production Editor: Angélique Gcouta
Printed on acid-free paper   SPIN: 10718346   27/3020GC 5 4 3 2 1 0

# Preface

The methylation of cytosine residues in DNA directly impacts human carcinogenesis in several ways. Firstly, methylation of cytosine markedly increases the probability of C to T transition mutations occurring at CpG sites. The coding regions of many tumor suppressor genes contain methylation sites, which are often hot spots for mutations both in the germ line and in somatic cells. Indeed, as many as 50% of all the transition mutations in the p53 gene in colon cancer occur at sites of cytosine methylation. In addition to its substantial effects on the occurrence of inactivating mutations, the methylation of cytosine residues in the promoters of tumor suppressor genes is also of major importance in the inactivation of these genes. Abnormal promoter methylation has been recognized as a third pathway to satisfy Knudson's hypothesis for the silencing of tumor suppressor genes in cancer. Important new advances have recently been made on the mechanisms underlying abnormal methylation in cancer and how the methylation signal is interpreted in the cell at the level of chromatin structure. This book presents the latest information regarding the role of cytosine methylation in the genetics and epigenetics of human cancer.

The opening chapter by Pfeifer, Tang, and Denissenko discusses mechanisms for increased mutagenesis at methylation sites induced by both endogenous and exogenous agents. The finding that cytosine methylation can increase the rates of mutations by enhancing the binding of chemical carcinogens to DNA was not previously realized, yet this is likely to have important implications for both chemical and ultra violet light induced carcinogenesis. Keith Robertson discusses the role of DNA methylation in modulating EBV gene expression in the second chapter. The life cycle of this virus is strongly influenced by alterations in methylation of the various promoters, an important finding being that methylation of a single site can result in substantial downregulation of viral genes. In the third chapter, James Herman and Steven Baylin summarize important experiments, which have implicated promoter hypermethylation and gene silencing in cancer development. There are now numerous examples of inappropriate promoter methylation in cancer discovered by these and other authors, and this work has covered a lot of ground in identifying promoter methylation as a frequent mechanism for tumor suppressor gene extinction. Brian Henrich and Adrian Bird in the fourth chapter discuss the mammalian methyl transferases and methylated DNA binding proteins (MDBs). This work has had major impact on our understanding of the mechanisms by which the epigenetic signal is interpreted in the cell and in explaining how chromatin structure is altered by changes in methylation patterns. In the fifth chapter Chan, Liang, and Jones discuss the relationship between transcription and DNA methylation, with the major emphasis being on the state of methylation of sites downstream of transcriptional initiation. This leads to a paradoxical situation, in which increased transcription through a region often is associated with increased methylation rather than with the inverse situation seen in the gene promoters. Andrew Feinberg discusses his pioneering work on the relationship between genomic imprinting and cancer in the sixth chapter. Loss of imprinting, as initially defined by Feinberg, has now been realized to be an important process by which imprinted genes contribute to carcinogenesis. Jean-Piene Issa focuses on methylation changes, which occur in apparently normal epithelium of older patients, in the next chapter. The finding of CpG island methylation in normal epithelium was entirely unexpected and is likely to have important implications for our understanding of the relationship between aging and cancer. Peter Laird outlines the use of mouse models in DNA methylation research in the eighth chapter and emphasizes the fact that knockout mice have had substantial impact on our understanding of the physiological role of methylation in normal development processes and in cancer. In the final chapter, Michael Lübbert discusses how DNA methylation inhibitors are finding increasing roles in the treatment of myelodysplastic syndromes, leukemias, and hemoglobinopathies. Clinical trials using methylation inhibitors have been quite promising and he points out the need to develop new allosteric inhibitors which affect methylation patterns so that abnormal methylation events can be reversed in preoplastic and neoplastic cells.

The importance of DNA methylation in human cancer has only become apparent over the last 5–10 years. This selective grouping of articles by some of the leading authorities in the field is therefore highly timely and will set the stage for future studies in which the prevalence of methylation defects in cancer will become apparent and new clinical strategies can be designed to alter this fundamental process in human cells.

Los Angeles and La Jolla
January 2000

PETER A. JONES
PETER K. VOGT

# List of Contents

# List of Contributors

(Their addresses can be found at the beginning of their respective chapters.)

BAYLIN, S.B.   35

BIRD, A.   55

CHAN, M.F.   75

DENISSENKO, M.F.   1

FEINBERG, A.P.   87

HENDRICH, B.   55

HERMAN, J.G.   35

ISSA, J.-P.   101

JONES, P.A.   75

LAIRD, P.W.   119

LIANG, G.   75

LÜBBERT, M.   135

PFEIFER, G.P.   1

ROBERTSON, K.D.   21

TANG, M.-S.   1

# Mutation Hotspots and DNA Methylation

G.P. Pfeifer[1], M.-s. Tang[2], and M.F. Denissenko[3]

## 1 Introduction

About 3–4% of all cytosines in mammalian DNA are methylated. Most or all of these 5-methylcytosine bases are found in the dinucleotide sequence CpG (RIGGS and JONES 1983; GRÜNWALD and PFEIFER 1989; BIRD 1992). As discussed in this chapter and elsewhere in this book, CpG methylation may play a critical role in carcinogenesis. Local increases or decreases in the extent of DNA methylation have been found in the DNA of tumor cells and have been correlated with tumor development. These methylation changes have been described as epigenetic events related to tumorigenesis. However, the hypermutability of CpG sequences, largely attributed to deamination of 5-methylcytosine, has been considered a source of genetic mutation in tumors (JONES et al. 1992; JONES 1996; LAIRD and JAENISCH 1996; GONZALGO and JONES 1997). 5-Methylcytosine was first identified as a mutational hotspot in *Escherichia coli* around two decades ago (COULONDRE et al. 1978). CpG sequences, which are presumably methylated, are preferentially mutated in many different human genetic diseases – for instance, in the factor IX gene

[1] Department of Biology, Beckman Research Institute of the City of Hope, Duarte, CA 91010, USA
E-mail: gpfeifer@smtplink.coh.org
[2] New York University, Dept. of Environmental Medicine, Tuxedo, NY 10987, USA
[3] PharMingen, San Diego, CA 92121, USA

in hemophilia (BOTTEMA et al. 1993; SOMMER 1995). In the *HPRT* gene of human lymphocytes, the most frequent mutational events in dividing cells in vivo (in somatic cells and in germ cells) are C-to-T substitutions at CpGs (O'NEILL and FINETTE 1998). These transitions are thought to result from endogenous deamination of 5-methylcytosine so that the methylated CpG dinucleotide is viewed as inherently mutagenic. This sequence may be subject to mutational changes occurring without DNA replication (GONZALGO and JONES 1997; RODIN and RODIN 1998). Methylation-mediated mutagenesis events apparently have had a strong enough impact that many CpG sites have been lost from the genome during evolution. In mammalian cells, CpGs are represented only at one-fifth of their expected frequency (SVED and BIRD 1990; BIRD 1992; SCHORDERET and GARTLER 1992). A normal frequency of CpGs is maintained only at CpG islands, sequences which probably are not methylated in the germ line (BIRD 1986).

Distinct DNA-methylation-mediated mechanisms may act under conditions of environmental exposure to exogenous carcinogens. Modification of DNA with chemical agents or exposure to $\gamma$ or ultraviolet (UV) irradiation reduces the ability of DNA to accept methyl groups (BOEHM and DRAHOVSKY 1983; WILSON and JONES 1983; PFEIFER et al. 1984; WACHSMAN 1997). This can lead to hypomethylation at CpG sites. A number of studies addressed this phenomenon in the early 1980s, but this line of research is not currently pursued very actively. Not only does carcinogen exposure affect methylation levels, but the opposite is also true: CpG methylation can have a strong influence on the formation of DNA lesions. Carcinogens like benzo[a]pyrene, found in tobacco smoke, have a much higher affinity towards methylated CpGs than towards their unmethylated counterparts (DENISSENKO et al. 1997; CHEN et al. 1998; PFEIFER and DENISSENKO 1998). The consequences of unrepaired DNA damage at CpGs (after replication and mutation fixation) may include mutations important for tumor formation. In this chapter, we will consider factors that may be responsible for the high mutation rates seen at CpG dinucleotides in tumor-suppressor genes, such as p53.

# 2 p53 Mutations in Tumors

Proto-oncogenes and tumor-suppressor genes may be critical selectable targets for mutations in tumors. The readiness with which CpG sequences undergo alteration will likely be involved in shaping mutational spectra in tumors. This became clear by an analysis of mutational events in the p53 tumor-suppressor gene. It has been estimated that more than 50% of all human tumors have a mutation in this gene (GREENBLATT et al. 1994) Thus, mutation in p53 is the single most common alteration in human cancers (HARRIS 1996), providing a unique opportunity to investigate the etiology, epidemiology, and pathogenesis of this disease (HOLLSTEIN et al. 1991; CARON DE FROMENTEL and SOUSSI 1992; GREENBLATT et al. 1994;

LEVINE et al. 1995; HOLLSTEIN et al. 1998; HUSSAIN and HARRIS 1998). More than 290 of the 393 codons of the p53 gene have been reported to harbor mutations, according to the July 1998 edition of the mutation database (HAINAUT et al. 1998). Most of these mutations are mis-sense mutations, and the latest database has more than 9000 entries and is still growing rapidly. The large number of entries adds more statistical power to distinct regularities of mutation distribution noted earlier (HOLLSTEIN et al. 1991; GREENBLATT et al. 1994). A notably high number of mutations are found at CpG dinucleotides. CpG sequences in the p53-coding sequence have been shown to be highly methylated in all human tissues so far examined (RIDEOUT et al. 1990; MAGEWU and JONES 1994; TORNALETTI and PFEIFER 1995). The 23 methylated CpGs constitute only about 8% of the central DNA-binding-domain sequence between codons 120 and 290 (LAIRD and JAENISCH 1996). However, about 33% of all mutations in this region occur at these sites. The vast majority of these alterations are transitions, and an even higher percentage of germline mutations (up to 60%) occurs at CpG sites in patients with Li-Fraumeni syndrome (LAIRD and JAENISCH 1996). Therefore, methylated CpG dinucleotides are the single most important mutational targets in p53. Five major p53 mutational hotspots, i.e., codons 175, 245, 248, 273, and 282, all contain methylated CpGs.

Human tumors of different histological origins display different patterns of p53 mutations. In colon cancer, transitions at CpGs account for almost 50% of all point mutations, but only 10% of liver or lung cancers contain such mutations (GREENBLATT et al. 1994). In contrast, in lung cancer, a majority of about 32% of the mutations are G-to-T transversions, most of them ascribed to a guanine on the non-transcribed DNA strand (HUSSAIN and HARRIS 1998). Transition mutations at CpG are much more frequent in almost all other internal cancers or in the germline. They have been linked to elevated deamination of endogenous 5-methylcytosine bases (JONES et al. 1992; JONES 1996; LAIRD and JAENISCH 1996; GONZALGO and JONES 1997). The frequency of transitions at CpGs in the p53 gene seems to correlate inversely with the extent of presumed DNA exposure to exogenous carcinogens (YANG et al. 1996a). In general, G-to-T transversions are frequent in lung cancers, liver cancers, and laryngeal tumors; in skin cancers, transition mutations are largely confined to dipyrimidine sequences (GREENBLATT et al. 1994). These differences in mutational profiles for different tumor types suggest that exogenous carcinogens are implicated in p53 mutagenesis. There are several examples of human cancers where a causative exogenous agent is either known or is strongly suspected. Solar UV light is involved in the induction of non-melanoma skin tumors basal cell and squamous cell carcinoma (SETLOW 1974; URBACH 1989). p53 Mutations in these human skin cancers bear C-to-T and CC-to-TT transition signatures (BRASH et al. 1991; DUMAZ et al. 1993; ZIEGLER et al. 1993), two types of base substitutions specifically induced by UV light in experimental systems (WOOD et al. 1984; MILLER 1985; PFEIFER 1997). Benzo[a]pyrene, which preferentially damages guanine bases and is a major mutagenic component of tobacco smoke, induces predominantly G-to-T transversions in murine tumors (RUGGERI et al. 1993). The percentage of G-to-T transversions in p53 is unusually high in human lung tumors diagnosed in smokers (GREENBLATT et al. 1994; HERNANDEZ-BOUS-SARD and HAINAUT 1998). Another example links hepatocellular carcinomas from certain areas of the world to a specific action of aflatoxin B1 (AFB1) on the p53 gene (PUSIEUX et al. 1991; AGUILAR et al. 1993).

Mutations in lung cancer and skin cancer, but not hepatocellular carcinoma, also cluster at CpG dinucleotides. The high transition-mutation rate at methylated CpGs in many cancers may be explained by the elevated susceptibility of these sites to spontaneous deamination (LAIRD and JAENISCH 1996; YANG et al. 1996a; see below), although other mechanisms are also conceivable. However, it is more difficult to find a sound explanation for the prevalence of transversions at methylated CpGs in carcinogen-specific tumors like cancer of the lung if one considers only endogenous sources of mutations in the form of 5-methylcytosine deamination. Interestingly, base changes characteristic of skin cancer, i.e. transitions at CC or TC dipyrimidine sequences, also show an association with methylated CpGs (TOMMASI et al. 1997). In this chapter, we will provide alternative explanations for the origin of CpG-associated mutations in these human tumors.

# 3 Deamination of 5-Methylcytosine

Endogenous deamination of 5-methylcytosine is viewed as the main source of the elevated rate of transitions at p53 mutational hotspots in human internal cancers, in which the most favored mutation is a conversion of C to T (GONZALGO and JONES 1997). Both cytosine and 5-methylcytosine are subject to deamination, resulting in emergence of uracil and thymine, respectively. Hydrolytic deamination occurs at cytosine in double-stranded DNA at a relatively slow rate, with a half-life of about 30,000 years at 37°C and pH 7.4 (FREDERICO et al. 1990; LINDAHL 1993; SHEN et al. 1994). The chemistry of cytosine deamination involves a hydroxyl-ion attack on cytosine bases protonated at the N3 position (FREDERICO et al. 1993). An alternative addition–elimination reaction seems less likely in vivo (LINDAHL 1993). Deamination of cytosine can be expedited in vitro under acidic conditions and by using chemicals, such as sodium bisulfite (WANG et al. 1980; FREDERICO et al. 1990). 5-Methylcytosine is resistant to bisulfite-induced deamination. However, methylation at the 5 position of the base ring facilitates spontaneous hydrolytic deamination by increasing the rate of hydrolysis (WANG et al. 1982; EHRLICH et al. 1986, 1990; LINDAHL 1993; SHEN et al. 1994). As a result, 5-methylcytosines are deaminated two to four times more rapidly than cytosines (EHRLICH et al. 1990; LINDAHL 1993; SHEN et al. 1994). For double-stranded DNA, the difference is 2.2-fold (SHEN et al. 1994). This twofold enhancement is probably not enough to account for the elevated mutation rate at methylated CpG dinucleotides. The mutational effect may be augmented by the difference in repair of the resulting two mismatches. The reason for the accumulation of transitions at methylated CpG may be a relatively inefficient repair of T/G mismatches vs U/G mismatches (SCHMUTTE et al. 1995; NEDDERMANN et al. 1996). Uracil is an abnormal DNA base and is recognized and excised efficiently by the ubiquitous enzyme uracil–DNA glycosylase (LINDAHL 1993). An apurinic/apyrimidinic endonuclease, a polymerase and a ligase coordinately correct the resulting abasic site. A mismatch-specific thymine–DNA glycosylase identified initially by Jiricny and coworkers (NEDDERMANN et al. 1996) was recently shown to remove etheno-cytosine residues from DNA with much higher efficiency than normal thymines from T/G mismatches (SAPARBAEV and LAVAL 1998; HANG et al. 1998). It is currently unclear whether this enzyme is the only activity that operates at T/G mismatches in vivo. Some of these mismatches may be corrected by the general mismatch-repair system (BILL et al. 1998), although it is unclear how strand specificity can be provided.

Mammalian homologues of the bacterial very-short-patch-repair (vsr)-gene product, which corrects T/G mismatches arising at dcm methylation sites through an endonucleolytic activity (SOHAIL et al. 1990; LIEB 1991; HENNECKE et al. 1991), have not yet been identified. Since the T/G mismatch is probably repaired less efficiently than a U/G mismatch, this consequently may create a higher risk for mutation fixation (LINDAHL 1993; SCHMUTTE et al. 1995). However, in an Alu element in p53 intron 6, the rate of CpG germ-line mutation in primate species was estimated to be about 1250 times slower than the in vitro deamination rate of 5-methylcytosine in double-stranded DNA (YANG et al. 1996b). The germ-line mutation rate was calculated to be even slower at CpGs in the factor-IX gene (SOMMER 1995). This implies that the existing repair mechanisms may correct both U/G and T/G mismatches quite efficiently or that deamination in vivo is much slower than deamination in vitro. In fact, it has not been proven beyond doubt that the deamination model accurately reflects mutagenesis at CpG sequences in mammalian cells. One major dilemma is the following: it has been calculated that only two 5-methylcytosines deaminate per day in each cell (SCHMUTTE and JONES 1998). These numbers appear almost insignificant compared with steady-state levels that have been measured for many endogenous and exogenous DNA adducts, which can be between hundreds and several hundred thousand per cell (MARNETT and BURCHAM 1993; HOLMQUIST 1998). It was estimated that each cell sustains up to 100,000 hits daily from oxidative DNA damage alone (WAGNER et al. 1992).

It is possible that certain chemicals may promote the deamination reactions at CpGs. Nitric oxide was shown to increase the rate of G/C-to-A/T transitions in Salmonella, perhaps via stimulation of deamination (WINK et al. 1991). Direct assays have failed to show any significant deamination of 5-methylcytosine by nitric oxide (SCHMUTTE et al. 1994; FELLEY-BOSCO et al. 1995). However, there is a relationship between the frequency of G/C-to-A/T transitions at CpGs in the p53 gene and increased nitric oxide synthase-2 expression in human colon carcinomas (AMBS et al. 1999). There are more examples of other mechanisms that may affect cytosine deamination directly or via formation of intermediates. 5-Methylcytosine can be deaminated by photochemical processes (PRIVAT and SOWERS 1996). Oxidative damage to 5-methylcytosine results primarily in formation of the deaminated product thymine glycol, which is primarily a replication-blocking lesion (which can, however, pair with adenine when bypassed by a polymerase; ZUO et al. 1995). Glyoxal, a known mutagen, was recently shown to directly deaminate

5-methylcytosine to thymine at a higher rate than it deaminated cytosine to uracil (KASAI et al. 1998). It has also been reported that ethylene oxide, a rodent and probable human carcinogen, and 1-nitropyrene, a common environmental mutagen, have the capacity to promote cytosine deamination (LI et al. 1992; MALIA and BASU 1994). Deamination was observed in DNA modified with mutagenic propylene oxide and acrylonitrile epoxide (SOLOMON and SEGAL 1989). Effects on deamination of 5-methylcytosine have not been measured for most of these compounds.

# 4 Enzyme-Mediated Mechanisms

An alternative pathway is believed to employ the intrinsic mutagenic capacity of the de novo methylation reaction at CpG sequences. Using in vitro systems, it has been demonstrated that several bacterial methyltransferases, including *Hpa*II, *Sss*I and others, promote C-to-U deaminations at CpG targets at low concentrations of the methyl-group donor S-adenosylmethionine (SHEN et al. 1992; WYSZYNSKI et al. 1994; YANG et al. 1995). The methyl-group transfer catalyzed by the *Hha*I methyltransferase was shown to occur through formation of the active intermediate between a cysteine residue of the enzyme and a cytosine base swung completely out of the DNA helix (KLIMASAUSKAS et al. 1995). The half-life of this intermediate may increase when the concentration of S-adenosylmethionine is low. This and the demonstrated higher methyltransferase affinity towards T/G and U/G mismatches than towards normal C/G base pairs (YANG et al. 1995; GONZALGO et al. 1997) may together provide an enzyme-mediated mechanism leading to the hypermutability of CpG dinucleotides. One bacterial methyltransferase was shown to convert 5-methylcytosine directly to thymine (YEBRA and BAGHWAT 1995). The proposal that enzyme-catalyzed events may play a role in carcinogenesis is supported by a number of studies reporting elevated expression of cytosine–DNA methyltransferase in human colon cancer cell lines and in colonic mucosa (EL-DEIRY et al. 1991; SCHMUTTE et al. 1996). This increase may reflect, at least in part, an increase in cell proliferation (LEE et al. 1996), with DNA methyltransferase being an S-phase-specific protein (VOGEL et al. 1988).

It is not clear, however, whether enzyme-mediated deamination is a significant event in vivo (WYSZYNSKI et al. 1994) or whether it can be carried out by mammalian DNA methyltransferases (SMITH et al. 1992). The extent of the involvement of enzyme-mediated deamination in CpG mutagenesis is not exactly defined (YANG et al. 1996a). Although *E. coli* DNA methyltransferase was able to cause deamination at the site of methylation, the in vivo experiments suggested that enzyme-catalyzed deaminations of cytosine did not play a major role in rendering methylation sites in *E. coli* hotspots for mutation (WYSZYNSKI et al. 1994). One study analyzed biochemical conditions that may favor the DNA-methyltransferase-mediated pathway in CpG mutagenesis (SCHMUTTE et al. 1996). The level of DNA-methyltransferase expression, the concentration of S-adenosylmethionine, the activity of uracil–DNA glycosylase (which may correct the resulting U base), and the presence of mutations in the methyltransferase gene itself were investigated. None of these parameters was significantly altered in colon-carcinoma tissues. Therefore, the mutagenic pathway C → U → T may not be the major factor responsible for the high rate of transition at CpGs in the p53 gene of colon cancers, but it may play a role under certain conditions (WYSZYNSKI et al. 1994; SCHMUTTE et al. 1996).

This raises the possibility that yet another endogenous or exogenous chemical mechanism may preferentially act at guanine bases in methylated CpG sites generating G-to-A transition mutations, which would be indistinguishable from C-to-T transitions. Such a postulate is further supported by observations that certain DNA alterations are strongly augmented at guanines in methylated CpG sequences.

## 5 Methylated CpG Sequences as Preferred Targets for Carcinogens

Mutations in the p53 gene are observed in about half of all human cancers, which reflects not only an important cellular function for this tumor suppressor but also suggests a variety of pathways through which the protein may be inactivated. Site-selective formation of DNA adducts, efficiency of repair, fidelity of replicative bypass of lesions (and, therefore, a stage of the cell cycle), and a complex interplay of tumor-selection mechanisms all may be involved in creating a clonal p53 mutation (Fig. 1). The fact that there are more than 200 codons that can be mutated in tumors raises the question of why mutational hotspots occur in this gene. In addition to the possibility of selection, the DNA damage spectrum or a sequence selectivity of repair may shape the p53 mutational spectrum. Our work has mainly focused on the first two steps of the mutagenesis process: the sequence-specific formation of DNA lesions and their repair rates along exons of the p53 gene in carcinogen-exposed human cells. We believe that many mutations affecting amino acids in the DNA-binding domain of the p53 protein (amino acids 120–290) are tumorigenic mutations (PFEIFER and DENISSENKO 1998). With the possible exception of hepatocellular carcinoma, selection may not be the dominant driving force

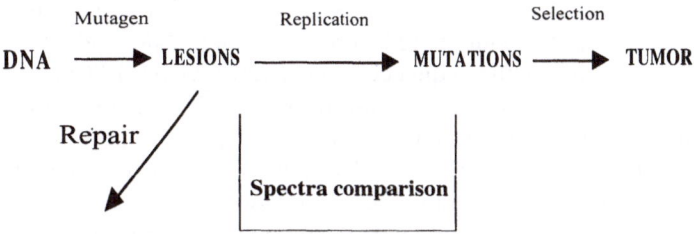

**Fig. 1.** Outline of the processes of mutagenesis that may lead to p53 mutations

that shapes the mutational spectra in tumors. Speculations about the interplay and priorities of these driving forces are beyond the scope of the present paper, and the reader is referred to additional articles (Soussi and May 1996; Pfeifer and Holmquist 1997; Denissenko and Pfeifer 1998; Strauss 1997; Rodin et al. 1998; Walker et al. 1999).

It is widely accepted that mutagenesis is an important component of tumorigenesis, a concept proven experimentally in animal models in which carcinogen exposure led to tumors harboring mutations in *ras* genes or in the *p53* gene. To determine the contribution of DNA damage to p53 mutational spectra, our approach involves identification of sequence-specific DNA lesions generated by direct-acting carcinogens in the *p53* gene and correlation of these DNA-damage "fingerprints" with p53 mutations collected from human cancer databases (Pfeifer and Denissenko 1998). This approach is based on mapping of DNA damage at nucleotide resolution by the ligation-mediated polymerase chain reaction (LMPCR) technique (Pfeifer et al. 1991, 1992, 1993). Using this technique, we have compared the distribution of DNA damage in the p53 gene of human cells exposed to UV light, benzo[a]pyrene diolepoxide (BPDE), or AFB1 with the distribution of p53 mutations in human cancers of the skin (non-melanoma), lung, and liver (Tornaletti and Pfeifer 1994; Denissenko et al. 1996; Tommasi et al. 1997; Denissenko et al. 1998a, 1998b). These experiments revealed a previously unrecognized role of methylated CpG sites as favored targets for a variety of physical and chemical genotoxic agents.

## 5.1 Role of 5-Methylcytosine-Associated DNA Damage in Skin Cancer

Exposure to solar radiation is a principal factor in the development of skin cancer (Mortimer 1991). Mutations in the p53 gene were found in many human non-melanoma skin malignancies (Brash et al. 1991; Dumaz et al. 1993; Ziegler et al. 1993) and in precursor lesions (Ziegler et al. 1994). Even normal sun-exposed skin contains a large number of clonal patches of p53-mutated keratinocytes (Jonason et al. 1996). The vast majority of base changes in skin lesions are C-to-T or CC-to-TT mutations at dipyrimidine sequences. These mutations are consistent with the specificity of the most mutagenic UV-induced lesion in mammalian cells, the cyclobutane–pyrimidine dimer (Pfeifer 1997). p53 Mutations in skin cancer are clustered at several mutational hotspots (Brash et al. 1991; Moles et al. 1993; Ziegler et al. 1994; Tommasi et al. 1997). With the exception of codons 177 and 278, all other skin-cancer hotspots (codons 152, 196, 213, 245, 248, and 282) contain the mutated dipyrimidine in the sequence context 5'CCG or 5'TCG. A review of a large number of publications on p53 mutations in skin tumors shows that the relative contribution of transition mutations affecting dipyrimidines within CpG sequences is 130/362 (36%). Remarkably, 5'CCG and 5'TCG occur only 19 times in the 1000-bp double-stranded target area between codons 120 and 290. All these CpG sequences are methylated in human keratinocytes (Tornaletti and Pfeifer 1995).

Using 254-nm UV-C light for irradiation, we found that only some of these skin-cancer hotspots were highly susceptible to UV damage formation (TORN-ALETTI et al. 1993). In particular, a lack of correlation was noted at dipyrimidine sequences that contain 5-methylcytosine. Subsequently, we found that mutation-hotspot positions that contain 5-methylcytosine within dipyrimidine sequences are up to 15-fold more susceptible to pyrimidine-dimer formation after exposure to natural sunlight (TOMMASI et al. 1997). Another study has reported a similar phenomenon in human cells irradiated with UV-B (280–320nm), a component of natural sunlight which reaches the Earth's surface (DROUIN and THERRIEN 1997). Methylation of cytosine enhances pyrimidine-dimer formation by sunlight 5–15-fold (TOMMASI et al. 1997). This difference may be explained by the higher energy absorption by 5-methylcytosine compared with cytosine in DNA. The $\lambda_{max}$ of 5-methylcytosine vs cytosine is red-shifted by about 6nm so that the $\lambda_{max}$ of 5-methylcytosine is 273nm compared with 267nm for cytosine at neutral pH. This red-shift results in a wavelength-dependent extinction coefficient that is 5–15 times higher for 5-methylcytosine than for cytosine at wavelengths from 300nm to 315nm (Fig. 2). This critical part of the solar spectrum is a component of sunlight that reaches the Earth's surface and is absorbed by DNA (SETLOW 1974). In addition, 5-methyl-deoxycytidine monophosphate (dCMP) has a significantly lower excited-singlet-state energy than thymidine monophosphate or dCMP (RUZCICSKA and LEMAIRE 1995) and, therefore, 5-methyl-dCMP may be a singlet energy trap in DNA. p53 Mutations induced by a solar UV simulator in the skin of hairless mice have been investigated (DUMAZ et al. 1997). It has been confirmed that the UV-B portion of the solar spectrum plays a major role in bringing about p53 alterations. The fact that sunlight induces cyclobutane–pyrimidine dimers preferentially at 5-methylcytosine bases was not recognized previously but could have important implications for sunlight-induced mutagenesis not limited to the p53 gene. More-

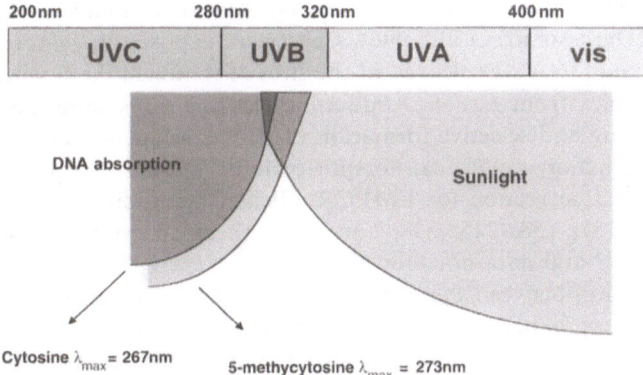

**Fig. 2.** Increased energy absorption of 5-methylcytosine versus cytosine between 300nm and 315nm, a critical part of the solar spectrum that reaches the Earth's surface and is thought to induce skin cancer. This difference could explain the existence of mutational hotspots at dipyrimidines that contain 5-methylcytosine

over, the 5-methylcytosine bases within pyrimidine dimers are prone to hydrolytic deamination, which probably increases their mutagenicity (DOUKI and CADET 1994; TU et al. 1998). These deamination reactions may be followed by a correct polymerase bypass of the deaminated dimers during DNA replication with incorporation of deoxyadenosine triphosphate (JIANG and TAYLOR 1993) and selective C- or 5-methyl-C-to-T transition mutations, the predominant types of mutations seen in non-melanoma skin tumors.

## 5.2 DNA Damage at Methylated CpG Sequences and Lung Cancer

Tobacco smoking is strongly associated with the development of lung cancer in humans (LOEB et al. 1984; HECHT et al. 1993). The majority of p53 mutations in the database are G-to-T transversions biased to guanines on the non-transcribed DNA strand (GREENBLATT et al. 1994; DENISSENKO et al. 1996; HERNANDEZ-BOUSSARD and HAINAUT 1998). Smoking-associated lung cancer is characterized by prominent mutational hotspots at codons 157, 158, 179, 245, 248, and 273, two of which (157 and 179) are common mutation sites only in tumors of the lung. Cancers from nonsmokers or from uranium miners do not contain these hotspot mutations (HERNANDEZ-BOUSSARD and HAINAUT 1998; PFEIFER and DENISSENKO 1998; PFEIFER et al. 1998) except for the A-to-G transition at codon 179, which is also a hotspot in nonsmokers. Such a peculiar mutation spectrum and signature suggest that distinct exogenous factors may be involved in lung tumorigenesis. G-to-T transversions are typical for bulky adduct-producing mutagens, including the class of polycyclic aromatic hydrocarbons (PAHs). Benzo[a]pyrene, a highly carcinogenic member of the PAH class, is present in quantities of 20–50ng per cigarette. Upon metabolic activation to BPDE, it induces mostly G-to-T mutations (EISENSTADT et al. 1982; CHEN et al. 1990) thought to result from covalent adducts formed at the N2 position of guanine. The distribution of BPDE adducts along the p53 gene was mapped at nucleotide resolution in carcinogen-treated normal human bronchial epithelial cells (DENISSENKO et al. 1996). Cells were exposed to various concentrations of BPDE, and DNA was cleaved at the sites of modified bases with the UvrABC nuclease complex from *E. coli*. Adduct-induced incisions were then visualized by LMPCR. Strong and selective formation of BPDE adducts occurred at guanine positions in the major mutational-hotspot codons (DENISSENKO et al. 1996). After quantitation and correction for LMPCR efficiency at individual nucleotide positions, codons 157, 158, 245, 248, and 273 appear to be the most strongly modified sites (Fig. 3 and data not shown).

There is a tight correlation between the benzo[a]pyrene-adduct spectrum and the mutation spectrum in lung cancer. This result provided a direct connection between carcinogens present in tobacco smoke and human cancer mutations. Subsequently, the mechanistic basis for the selective occurrence of these damage hotspots was shown to be related to patterns of cytosine methylation in the p53 gene (DENISSENKO et al. 1997). The distribution of BPDE–DNA adducts differed drastically in CpG-methylated DNA compared with non-methylated DNA. These

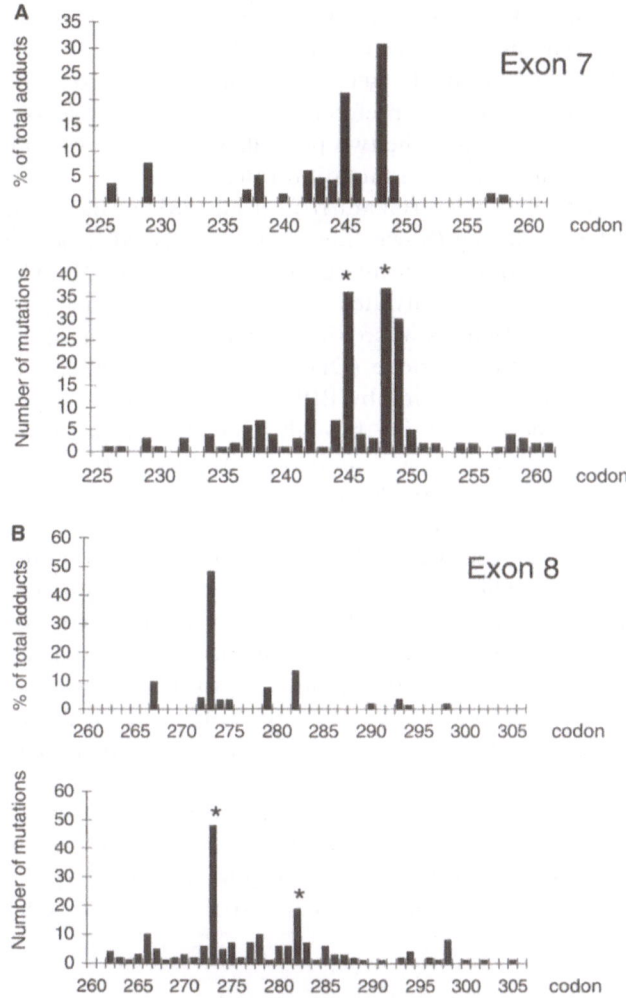

**Fig. 3.** Mapping and quantitation of benzo[a]pyrene diolepoxide (BPDE) DNA adducts along exons 7 and 8 of the human p53 gene and correlation with mutations in smoking-induced lung cancers. **A** Exon 7. **B** Exon 8. The p53-mutation database was used for determining the frequency distribution of mutations. Cancers from non-smokers and from uranium miners were excluded. Lung cancers from non-smokers contain a mutational hotspot at codon 179 only, and miners' cancers display a hotspot at codon 249 (data not shown). BPDE DNA adducts in the p53 gene of human bronchial epithelial cells were mapped as described (DENISSENKO et al. 1996). The data were quantified by phosphorimaging and by correcting for ligation-mediated polymerase-chain-reaction amplification efficiency at individual sites (DENISSENKO et al. 1998b). The *asterisks* indicate mutation hotspots at methylated CpG sequences

results are not due to a preference of the UvrABC nuclease for methylated sequences (TANG et al. 1999). Guanines flanked by 5-methylcytosines were the preferentially adducted positions. Methylation enhanced adduct formation by three- to tenfold, depending on the sequence. This tendency was confirmed in genomic DNA, where adducts were measured in human *PGK1* genes differing in methylation (DENISSENKO et al. 1997). Therefore, CpG dinucleotides that are methylated in the human p53 gene in all tissues examined (RIDEOUT et al. 1990; MAGEWU and JONES 1994; TORNALETTI and PFEIFER 1995), in addition to being endogenous pro-mutagenic factors, represent preferential targets for exogenous chemical carcinogens as well. Figure 4 shows the two possible pathways operating at methylated CpG sites and creating increased mutation rates.

In the case of BPDE, hydrophobic effects (GEACINTOV et al. 1988) or increased molecular polarizability and base stacking (SOWERS et al. 1987) derived from the methyl group may facilitate the creation of an intercalation site for BPDE. The increase in BPDE intercalative binding to methylated CpG sites may eventually be reflected in the extent of covalent interactions where the benzo[a]pyrene ring of the (+) adduct is associated with the minor groove (DENISSENKO et al. 1997). The precise mechanism of methylated CpG targeting by BPDE is a subject of further studies. Likewise, other DNA adducts that arise through intercalation may form more easily at methylated CpGs. Recently, several bulky chemical carcinogens (benzo[g]chrysene diol epoxide, AFB1 8,9-epoxide, and N-acetoxy-2-aminofluorene) were shown to bind preferentially at methylated CpGs in the p53 gene (CHEN et al. 1998). Importantly, these carcinogens react with different moieties within the guanine base structure (CHEN et al. 1998). Mitomycin cross-linking is slightly enhanced by methylation (MILLARD and BEACHY 1993; JOHNSON et al. 1995). Nitrosopyrene and mitomycin monoadducts are greatly enhanced at methylated CpGs (Tang et al., unpublished results). In other studies, cytosine methylation increased DNA damage produced by intercalating and groove-binding enediynes (MATHUR et al. 1997). This suggests that a mechanism more universal than intercalation may operate at methylated CpGs.

The extent to which enhanced binding of an individual carcinogen at methylated CpG affects mutagenesis at the same location has not yet been studied but is expected to increase CpG mutation frequencies. The mutagenic responses of the endogenous *HPRT* gene and a *lacI* transgene were assessed in a mouse model

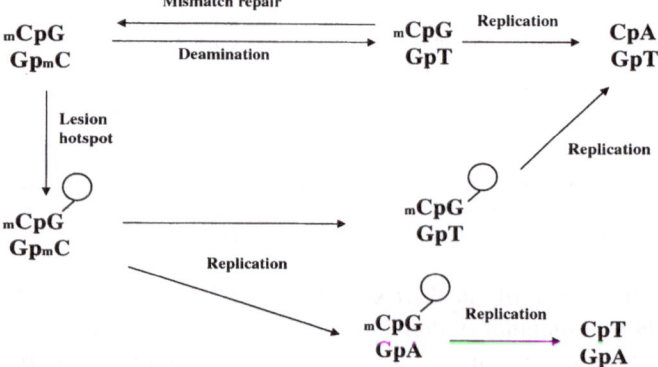

**Fig. 4.** Possible mechanisms that may operate at methylated CpG sequences to produce mutational hotspots

system following exposure to benzo[a]pyrene (SKOPEK et al. 1996). This exposure produced many more mutations at the *lacI* gene than at the *HPRT* locus. The fact that the former is rich in CpG sequences and the latter contains relatively few CpGs may explain these significant differences (SKOPEK et al. 1996). In fact, almost all CpG sites in the *lacI* gene of transgenic mice are methylated in all tissues analyzed (YOU et al. 1998). The spontaneous mutation rate of the *lacI* transgene is rather high, presumably due to increased deamination of these 5-methylcytosines or other methylation-mediated mechanisms. Methylated CpG sequences in the *lacI* gene may also be preferential binding sites for certain mutagens. These findings further substantiate the role of CpG methylation in the mutagenic response of an individual gene. Therefore, the frequencies and methylation status of CpG sites within a gene target need to be determined in order to draw meaningful conclusions about mutational spectra.

Another crucial component of the mutagenesis process is DNA repair. It was suggested that the strand-specific distribution of CpG transition mutations in the p53 gene is essentially random, and the strand bias at certain codon positions may have been due to other confounding factors, including mutant selection (YANG et al. 1996a; Rodin et al. 1998). In contrast, G-to-T base changes at CpG sequences exhibit a profound strand bias ($> 90\%$) towards the non-transcribed strand in lung, breast, head and neck, and liver cancers (HUSSAIN and HARRIS 1998). This implies a strand-specific DNA-repair phenomenon in which bulky DNA adducts on the non-transcribed strand are repaired less efficiently than those on the transcribed strand (MELLON et al. 1987; CHEN et al. 1992). Repair experiments analyzing BPDE adducts in the p53 gene have shown that the non-transcribed strand is indeed repaired more slowly than the transcribed strand (DENISSENKO et al. 1998a). Notably, BPDE adducts were initially formed at a similar extent on both DNA strands, but the damage was removed significantly more slowly from methylated CpG hotspots positioned on the non-transcribed strand (DENISSENKO et al. 1998a). These observations demonstrate that repair efficiency may contribute to mutational spectra in human genes. The presence of a strong strand bias in the distribution of G-to-T transversions at CpG hotspot codons and the parallel-repair strand bias lend additional support to the proposal that lung cancer mutations are caused by PAHs in cigarette smoke.

The presence of 5-methylcytosine may affect the sequence-dependent repair of DNA adducts. A 5-methylcytosine adjacent in the 5'-position to an O6-methylated guanine strongly diminished repair of this lesion by O6-methylguanine-DNA methyltransferase (BENTIVEGNA and BRESNICK 1994). Most pyrimidine dimers containing a 5-methylcytosine as the dimerized base were repaired more slowly than neighboring sequences without 5-methylcytosine, and this correlated with the position of mutation hotspots in skin tumors (TORNALETTI and PFEIFER 1994). Further work is required to investigate more directly the influence of methylation on repair of DNA lesions at CpG sites by base-excision repair, mismatch repair, and nucleotide-excision repair complexes.

# 6 Conclusions

Mutations in the p53 gene are tumor-specific (GREENBLATT et al. 1994; HUSSAIN and HARRIS 1998). A high rate of transition mutations at CpG dinucleotides is found in colon carcinomas (up to 50%) and other internal cancers. Such a mutation spectrum is believed to result from enhanced spontaneous deamination of 5-methylcytosine. Unlike colon cancer, in lung cancers, G-to-T transversions at methylated CpGs are frequent and, in skin cancers, transitions at dipyrimidine sequences are the main mutations seen. This specificity suggests exogenous causative agents for certain tumors (GREENBLATT et al. 1994; HUSSAIN and HARRIS 1998; PFEIFER and DENISSENKO 1998). The unique palindromic structure of a methylated CpG site creates an additional possible source of a C-to-T mutational event. The observed mutations may be caused by a lesion residing preferentially at guanine bases in methylated CpG sequences, which would produce G-to-A transition mutations indistinguishable from C-to-T mutations on the opposite strand. This notion is supported by the finding that formation of many DNA lesions is strongly enhanced by methylation of CpG sites (DENISSENKO et al. 1997; TOMMASI et al. 1997, CHEN et al. 1998).

It is clear that 5-methylcytosine is a major player in mammalian mutagenesis. Its importance encompasses, but may not be limited to, at least two main pathways: enhanced spontaneous deamination and increased affinity of methylated CpG sites for DNA-reactive molecules.

*Acknowledgements.* The work of the authors was supported by National Institutes of Health grants CA65652 and CA69449 (to G.P.P.) and ES03124 and ES08389 (to M.S.T.).

# References

Aguilar F, Hussain SP, Cerutti P (1993) Aflatoxin B1 induces the transversion of G → T in codon 249 of the p53 tumor suppressor gene in human hepatocytes. Proc Natl Acad Sci USA 90:8586–8590

Ambs S, Bennett WP, Merriam WG, Ogunfusika MO, Oser SM, Harrington AM, Shields PG, Felley-Bosco E, Hussain SP, Harris CC (1999) Relationship between p53 mutations and inducible nitric oxide synthase expression in human colorectal cancer. J Natl Cancer Inst 91:86–88

Bentivegna SS, Bresnick E (1994) Inhibition of human O6-methylguanine-DNA methyltransferase by 5-methylcytosine. Cancer Res 54:327–329

Bill CA, Duran WA, Miselis NR, Nickoloff JA (1998) Efficient repair of all types of single-base mismatches in recombination intermediates in Chinese hamster ovary cells: competition between long-patch and G–T glycosylase-mediated repair of G–T mismatches. Genetics 149:1935–1943

Bird AP (1986) CpG-rich islands and the function of DNA methylation. Nature (London) 321:209–213

Bird AP (1992) The essentials of DNA methylation. Cell 70:5–8

Boehm TLJ, Drahovsky D (1983) Alteration of enzymatic DNA methylation by chemical carcinogens. Recent Results Cancer Res 84:212–225

Bottema CD, Ketterling RP, Vielhaber E, Yoon HS, Gostout B, Jacobson DP, Shapiro A, Sommer SS (1993) The pattern of spontaneous germ-line mutation: relative rates of mutation at or near CpG dinucleotides in the factor IX gene. Hum Genet 91:496–503

Brash DE, Rudolph JA, Simon JA, Lin A, McKenna GJ, Baden HP, Halperin AJ, Pontén J (1991) A role for sunlight in skin cancer: UV-induced p53 mutations in squamous cell carcinoma. Proc Natl Acad Sci USA 88:10124–10128

Caron de Fromentel C, Soussi T (1992) TP53 tumor suppressor gene: a model for investigating human mutagenesis. Genes Chromosomes Cancer 4:1–15

Chen JX, Zheng Y, West M, Tang M-S (1998) Carcinogens preferentially bind at methylated CpG in the p53 mutational hotspots. Cancer Res 58:2070–2075

Chen R-H, Maher VM, McCormick JJ (1990) Effect of excision repair by diploid human fibroblasts on the kinds and locations of mutations induced by (+)-7b,8a-dihydroxy-9a,10a-epoxy-7,8,9,10-tetra-hydrobenzo[a]pyrene in the coding region of the HPRT gene. Proc Natl Acad Sci USA 87:8680–8684

Chen R-H, Maher VM, Brouwer J, Van de Putte P, McCormick JJ (1992) Preferential repair and strand-specific repair of benzo(a)pyrene diol epoxide adducts in the HPRT gene of diploid human fibroblasts. Proc Natl Acad Sci USA 89:5413–5417

Coulondre C, Miller JH, Farabaugh PJ, Gilbert W (1978) Molecular basis of base substitution hotspots in *Escherichia coli*. Nature 274:775–780

Denissenko MF, Pao A, Tang M-S, Pfeifer GP (1996) Preferential formation of benzo[a]pyrene adducts at lung cancer mutational hotspots in P53. Science 274:430–432

Denissenko MF, Chen JX, Tang M-S, Pfeifer GP (1997) Cytosine methylation determines hot spots of DNA damage in the human P53 gene. Proc Natl Acad Sci USA 94:3893–3898

Denissenko MF, Pao A, Pfeifer GP, Tang M-S (1998a) Slow repair of bulky DNA adducts along the non-transcribed strand of the human p53 gene may explain the strand bias of transversion mutations in cancers. Oncogene 16:1241–1247

Denissenko MF, Koudriakova TB, Smith L, O'Connor TR, Riggs AD, Pfeifer GP (1998b) The P53 codon 249 mutational hotspot in hepatocellular carcinoma is not related to selective formation or persistence of aflatoxin B1 adducts. Oncogene 17:3007–3013

Douki T, Cadet J (1994) Formation of cyclobutane dimers and (6–4) photoproducts upon far-UV photolysis of 5-methylcytosine-containing dinucleoside monophosphates. Biochemistry 33:11942–11950

Drouin R, Therrien J-P (1997) UVB-induced cyclobutane pyrimidine dimer frequency correlates with skin cancer mutational hotspots in p53. Photochem Photobiol 66:719–726

Dumaz N, Drougard C, Sarasin A, Daya-Grosjean L (1993) Specific UV-induced mutation spectrum in the p53 gene of skin tumors from DNA-repair-deficient xeroderma pigmentosum patients. Proc Natl Acad Sci USA 90:10529–10533

Dumaz N, Van Kranen HJ, De Vries A, Berg RJ, Wester PW, van Kreijl CF, Sarasin A, Daya-Grosjean L, De Gruijl FR (1997) The role of UV-B light in skin carcinogenesis through the analysis of p53 mutations in squamous cell carcinomas of hairless mice. Carcinogenesis 18:897–904

Ehrlich M, Norris KF, Wang RY, Kuo KC, Gehrke CW (1986) DNA cytosine methylation and heat-induced deamination. Biosci Rep 6:387–393

Ehrlich M, Zhang X-Y, Inamdar NM (1990) Spontaneous deamination of cytosine and 5-methylcytosine residues in DNA and replacement of 5-methylcytosine residues with cytosine residues. Mutation Res 238:277–286

Eisenstadt E, Warren AJ, Porter J, Atkins D, Miller JH (1982) Carcinogenic epoxides of benzo(a)pyrene and cyclopenta(cd)pyrene induce base substitutions via specific transversions. Proc Natl Acad Sci USA 79:1945–1949

El-Deiry WS, Nelkin BD, Celano P, Yen RW, Falco JP, Hamilton SR, Baylin SB (1991) High expression of the DNA methyltransferase gene characterizes human neoplastic cells and progression stages of colon cancer. Proc Natl Acad Sci USA 88:3470–3474

Felley-Bosco E, Mirkovitch J, Ambs S, Mace K, Pfeifer A, Keefer LK, Harris CC (1995) Nitric oxide and ethylnitrosourea: relative mutagenicity in the p53 tumor suppressor and hypoxanthine-phosphor-ibosyltransferase genes. Carcinogenesis 16:2069–2074

Frederico LA, Kunkel TA, Shaw BR (1990) A sensitive genetic assay for the detection of cytosine deamination: determination of rate constants and the activation energy. Biochemistry 29:2532–2537

Frederico LA, Kunkel TA, Ramsey-Shaw B (1993) Cytosine deamination in mismatched base pairs. Biochemistry 32:6523–6530

Geacintov NE, Shabaz M, Ibanez V, Moussaoui K, Harvey RG (1988) Base-sequence dependence of non-covalent complex formation and reactivity of benzo[a]pyrene diol epoxide with polynucleotides. Biochemistry 27:8380–8387

Gonzalgo ML, Jones PA (1997) Mutagenic and epigenetic effects of DNA methylation. Mutation Res 386:107–118

Greenblatt MS, Bennett WP, Hollstein M, Harris CC (1994) Mutations in the p53 tumor-suppressor gene: clues to cancer etiology and molecular pathogenesis. Cancer Res 54:4855–4878

Grünwald S, Pfeifer GP (1989) Enzymatic DNA methylation. Prog Clin Biochem Med 9:61–103

Hainaut P, Hernandez T, Robinson A, Rodriguez-Tome P, Flores T, Hollstein M, Harris CC, Montesano R (1998) IARC database of p53 gene mutations in human tumors and cell lines: updated compilation, revised formats and new visualisation tools. Nucleic Acids Res 26:205–213

Hang B, Medina M, Fraenkel-Conrat H, Singer B (1998) A 55-kDa protein isolated from human cells shows DNA glycosylase activity toward 3,N4-ethenocytosine and the G/T mismatch. Proc Natl Acad Sci USA 95:13561–13566

Harris CC (1996) p53 Tumor suppressor gene: from the basic research laboratory to the clinic – an abridged historical perspective. Carcinogenesis 17:1187–1198

Hecht SS, Carmella SG, Murphy SE, Foiles PG, Chung F-L (1993) Carcinogen biomarkers related to smoking and upper aerodigestive tract cancer. J Cell Biochem Suppl 17F:27–35

Hennecke F, Kolmar H, Brundl K, Fritz HJ (1991) The vsr gene product of *E. coli* K-12 is a strand- and sequence-specific DNA mismatch endonuclease. Nature 353:776–778

Hernandez-Boussard TM, Hainaut P (1998) A specific spectrum of p53 mutations in lung cancer from smokers: review of mutations compiled in the IARC p53 database. Environ Health Perspect 106: 385–391

Hollstein M, Sidransky D, Vogelstein B, Harris CC (1991) p53 Mutations in human cancers. Science 253:49–53

Hollstein M, Moeckel G, Hergenhahn M, Spiegelhalder B, Keil M, Werle-Schneider G, Bartsch H, Brickmann J (1998) On the origins of tumor mutations in cancer genes: insights from the p53 gene. Mutation Res 405:145–154

Holmquist GP (1998) Endogenous lesions, S-phase-independent spontaneous mutations, and evolutionary strategies for base excision repair. Mutat Res 400:59–68

Hussain SP, Harris CC (1998) Molecular epidemiology of human cancer: contribution of mutation spectra studies of tumor suppressor genes. Cancer Res 58:4023–4037

Jiang N, Taylor J-S (1993) In vivo evidence that UV-induced C–T mutations at dipyrimidine sites could result from the replicative bypass of *cis-syn* cyclobutane dimers or their deamination products. Biochemistry 32:472–481

Johnson WS, He Q-Y, Tomasz M (1995) Selective recognition of the m5CpG dinucleotide sequence in DNA by mitomycin C for alkylation and cross-linking. Bioorg Med Chem 3:851–860

Jonason AS, Kunala S, Price GJ, Restifo RJ, Spinelli HM, Persing JA, Leffell DJ, Tarone RE, Brash DE (1996) Frequent clones of p53-mutated keratinocytes in normal human skin. Proc Natl Acad Sci USA 93:14025–14029

Jones PA (1996) DNA methylation errors and cancer. Cancer Res 56:2463–2467

Jones PA, Rideout WM, Shen J-C, Spruck CH, Tsai YC (1992) Methylation, mutation and cancer. BioEssays 14:33–36

Kasai H, Iwamoto-Tanaka N, Fukada S (1998) DNA modifications by the mutagen glyoxal: adduction to G and C, deamination of C and GC and GA cross-linking. Carcinogenesis 19:1459–1465

Klimasauskas S, Kumar S, Roberts RJ, Cheng X (1994) HhaI methyltransferase flips its target base out of the DNA helix. Cell 76:357–369

Laird PW, Jaenisch R (1996) The role of DNA methylation in cancer genetics and epigenetics. Annu Rev Genet 30:441–464

Lee PJ, Washer LL, Law DJ, Boland CR, Horon IL, Feinberg AP (1996) Limited up-regulation of DNA methyltransferase in human colon cancer reflecting increased cell proliferation. Proc Natl Acad Sci USA 93:10366–10370

Levine AJ, Wu MC, Chang A, Silver A, Attiyeh EF, Lin J, Epstein CB (1995) The spectrum of mutations at the p53 locus. Ann NY Acad Sci 768:111–128

Li F, Segal A, Solomon JJ (1992) In vitro reaction of ethylene oxide with DNA and characterization of DNA adducts. Chem Biol Interact 83:35–54

Lieb M (1991) Spontaneous mutation at a 5-methylcytosine hotspot is prevented by very short patch (VSP) mismatch repair. Genetics 128:23–27

Lindahl T (1993) Instability and decay of the primary structure of DNA. Nature 362:709–715

Loeb LA, Ernster VL, Warner KE, Abbotts J, Laszlo J (1984) Smoking and lung cancer: an overview. Cancer Res 44:5940–5958

Magewu AN, Jones PA (1994) Ubiquitous and tenacious methylation of the CpG site in codon 248 of the p53 gene may explain its frequent appearance as a mutational hot spot in human cancer. Mol Cell Biol 14:4225–4232

Malia SA, Basu AK (1994) Reductive metabolism of 1-nitropyrene accompanies deamination of cytosine. Chem Res Toxicol 7:823–828

Marnett LJ, Burcham PC (1993) Endogenous DNA adducts: potential and paradox. Chem Res Toxicol 6:771–785

Mathur P, Xu J, Dedon PC (1997) Cytosine methylation enhances DNA damage produced by groove binding and intercalating enediynes: studies with esperamicins A1 and C. Biochemistry 36:14868–14873

Mellon I, Spivak G, Hanawalt PC (1987) Selective removal of transcription blocking DNA damage from the transcribed strand of the mammalian DHFR gene. Cell 51:241–249

Millard JT, Beachy TM (1993) Cytosine methylation enhances mitomycin-C cross-linking. Biochemistry 32:12850–12856

Miller JH (1985) Mutagenic specificity of ultraviolet light. J Mol Biol 182:45–68

Molès JP, Moyret C, Guillot B, Jeanteur P, Guilhou JJ, Theillet C, Basset-Sèguin N (1993) p53 mutations in human epithelial skin cancers. Oncogene 8:583–588

Mortimer P (1991) Squamous cell and basal cell skin carcinoma and rarer histologic types of skin cancer. Curr Opin Oncol 3:349–354

Neddermann P, Gallinari P, Lettieri T, Schmid D, Truong O, Hsuan JJ, Wiebauer K, Jiricny J (1996) Cloning and expression of human G/T mismatch-specific thymine–DNA glycosylase. J Biol Chem 271:12767–12774

O'Neill JP, Finette BA (1998) Transition mutations at CpG dinucleotides are the most frequent in vivo spontaneous single-base substitution mutation in the human HPRT gene. Environ Mol Mutagen 32:188–191

Pfeifer GP (1997) Formation and processing of UV photoproducts: effects of DNA sequence and chromatin environment. Photochem Photobiol 65:270–283

Pfeifer GP, Denissenko MF (1998) Formation and repair of DNA adducts in the p53 gene: relation to cancer mutations? Environ Mol Mutagen 31:197–205

Pfeifer GP, Holmquist GP (1997) Mutagenesis in the p53 gene. Biochim Biophys Acta 1333:M1–M8

Pfeifer GP, Grunberger D, Drahovsky D (1984) Impaired enzymatic methylation of BPDE-modified DNA. Carcinogenesis 5:931–934

Pfeifer GP, Drouin R, Riggs AD, Holmquist GP (1991) In vivo mapping of a DNA adduct at nucleotide resolution: detection of pyrimidine (6–4) pyrimidone photoproducts by ligation-mediated polymerase chain reaction. Proc Natl Acad Sci USA 88:1374–1378

Pfeifer GP, Drouin R, Riggs AD, Holmquist GP (1992) Binding of transcription factors creates hot spots for UV photoproducts in vivo. Mol Cell Biol 12:1798–1804

Pfeifer GP, Drouin R, Holmquist GP (1993) Detection of DNA adducts at the DNA sequence level by ligation-mediated PCR. Mutat Res 288:39–46

Pfeifer GP, Denissenko MF, Tang M-S (1998) p53 Mutations, benzo[a]pyrene and lung cancer. Mutagenesis 13:537–538

Privat E, Sowers LC (1996) Photochemical deamination and demethylation of 5-methylcytosine. Chem Res Toxicol 9:745–750

Puisieux A, Lim S, Groopman J, Ozturk M (1991) Selective targeting of p53 gene mutational hotspots in human cancers by etiologically defined carcinogens. Cancer Res 51:6185–6189

Rideout III WM, Coetzee GA, Olumi AF, Jones PA (1990) 5-Methylcytosine as an endogenous mutagen in the human LDL receptor and p53 genes. Science 249:1288–1290

Riggs AD, Jones PA (1983) 5-Methylcytosine, gene regulation, and cancer. Adv Cancer Res 40:1–30

Rodin SN, Rodin AS (1998) Strand asymmetry of CpG transitions as indicator of G1 phase-dependent origin of multiple tumorigenic p53 mutations in stem cells. Proc Natl Acad Sci USA 95:11927–11932

Rodin SN, Holmquist GP, Rodin AS (1998) CpG transition strand asymmetry and hitch-hiking mutations as measures of tumorigenic selection in shaping the p53 mutational spectrum. Int J Mol Med 1:191–199

Ruggeri B, DiRado M, Zhang SY, Bauer B, Goodrow T, Klein-Szanto AJP (1993) Benzo[a]pyrene-induced murine skin tumors exhibit frequent and characteristic G to T mutations in the p53 gene. Proc Natl Acad Sci USA 90:1013–1017

Ruzcicska BP, Lemaire DGE (1995) DNA photochemistry. In: Horspool WM, Song P-S (eds) CRC handbook of organic photochemistry and photobiology. CRC Press, Boca Raton, Florida, pp 1289–1317

Saparbaev M, Laval J (1998) 3,N4-ethenocytosine, a highly mutagenic adduct, is a primary substrate for Escherichia coli double-stranded uracil–DNA glycosylase and human mismatch-specific thymine-DNA glycosylase. Proc Natl Acad Sci USA 95:8508–8513

Schmutte C, Jones PA (1998) Involvement of DNA methylation in human carcinogenesis. Biol Chem 379:377–388

Schmutte C, Rideout WM, Shen JC, Jones PA (1994) Mutagenicity of nitric oxide is not caused by deamination of cytosine or 5-methylcytosine in double-stranded DNA. Carcinogenesis 15:2899–2903

Schmutte C, Yang AS, Beart RW, Jones PA (1995) Base excision repair of U:G mismatches at a mutational hotspot in the p53 gene is more efficient than base excision repair of T:G mismatches in extracts of human colon tumors. Cancer Res 55:3742–3746

Schmutte C, Yang AS, Nguyen T-DT, Beart RW, Jones PA (1996) Mechanisms for the involvement of DNA methylation in colon carcinogenesis. Cancer Res 56:2375–2381

Schorderet DF, Gartler SM (1992) Analysis of CpG suppression in methylated and non-methylated species. Proc Natl Acad Sci USA 89:957–961

Setlow RB (1974) The wavelengths in sunlight effective in producing skin cancer: a theoretical analysis. Proc Natl Acad Sci USA 71:3363–3366

Shen JC, Rideout III WM, Jones PA (1992) High frequency mutagenesis by a DNA methyltransferase. Cell 71:1073–1080

Shen JC, Rideout III WM, Jones PA (1994) The rate of hydrolytic deamination of 5-methylcytosine in double-stranded DNA. Nucleic Acids Res 22:972–976

Skopek TR, Kort KL, Marino DR, Mittal LV, Umbenhauer DR, Laws GM, Adams SP (1996) Mutagenic response of the endogenous hprt gene and lacI transgene in benzo[a]pyrene-treated Big Blue B6C3F1 mice. Environ Mol Mutagen 28:376–384

Smith SS, Kaplan BE, Sowers LC, Newman EM (1992) Mechanism of human methyl-directed DNA methyltransferase and fidelity of cytosine methylation. Proc Natl Acad Sci USA 89:4744–4748

Sohail A, Lieb M, Dar M, Bhagwat AS (1990) A gene required for very short patch repair in Escherichia coli is adjacent to the DNA cytosine methylase gene. J Bacteriol 172:4214–4221

Solomon JJ, Segal A (1989) DNA adducts of propylene oxide and acrylonitrile epoxide: hydrolytic deamination of 3-alkyl-dCyd to 3-alkyl-dUrd. Environ Health Perspect 81:19–22

Sommer SS (1995) Recent human germ-line mutation: inferences from patients with hemophilia B. Trends Genet 11:141–147

Soussi T, May P (1996) Structural aspects of the p53 protein in relation to gene evolution: a second look. J Mol Biol 260:623–637

Sowers LC, Ramsey-Shaw B, Sedwick WD (1987) Base stacking and molecular polarizability: effect of a methyl group in the 5-position of pyrimidines. Biochem Biophys Res Commun 148:790–794

Strauss BS (1997) Silent and multiple mutations in p53 and the question of the hypermutability of tumors. Carcinogenesis 18:1445–1452

Sved J, Bird A (1990) The expected equilibrium of the CpG dinucleotide in vertebrate genomes under a mutational model. Proc Natl Acad Sci USA 87:4692–4696

Tang M-S, Zheng JB, Denissenko MF, Pfeifer GP, Zheng Y (1999) Use of UvrABC nuclease to quantify benzo[a]pyrene diol epoxide–DNA adduct formation at methylated versus unmethylated CpG sites in the p53 gene. Carcinogenesis 20:1085–1089

Tommasi S, Denissenko MF, Pfeifer GP (1997) Sunlight induces pyrimidine dimers preferentially at 5-methylcytosine bases. Cancer Res 57:4727–4730

Tornaletti S, Pfeifer GP (1994) Slow repair of pyrimidine dimers at p53 mutation hotspots in skin cancer. Science 263:1436–1438

Tornaletti S, Pfeifer GP (1995) Complete and tissue-independent methylation of CpG sites in the p53 gene: implications for mutations in human cancers. Oncogene 10:1493–1499

Tornaletti S, Rozek D, Pfeifer GP (1993) The distribution of UV photoproducts along the human p53 gene and its relation to mutations in skin cancer. Oncogene 8:2051–2057

Tu Y, Dammann R, Pfeifer GP (1998) Sequence and time-dependent deamination of cytosine bases in UVB-induced cyclobutane pyrimidine dimers in vivo. J Mol Biol 284:297–311

Urbach F (1989) Potential effects of altered solar ultraviolet radiation on human skin cancer. Photochem Photobiol 50:507–513

Vogel MC, Papadopoulos T, Müller-Hermelink HK, Drahovsky D, Pfeifer GP (1988) Intracellular distribution of DNA methyltransferase during the cell cycle. FEBS Lett 236:9–13

Waohaman JT (1997) DNA methylation and the association between genetic and epigenetic changes: relation to carcinogenesis. Mutation Res 375:1–8

Wagner J, R, Hu CC, Ames BN (1992) Endogenous oxidative damage of deoxycytidine in DNA. Proc Natl Acad Sci USA 89:3380–3384

Walker DR, Bond JP, Tarone RE, Harris CC, Makalowski W, Boguski MS, Greenblatt MS (1999) Evolutionary conservation and somatic mutation hotspot maps of p53: correlation with p53 protein structural and functional features. Oncogene 18:211–218

Wang RY, Gehrke CW, Ehrlich M (1980) Comparison of bisulfite modification of 5-methyldeoxycytidine and deoxycytidine residues. Nucleic Acids Res 8:4777–4790

Wang RY, Kuo KC, Gehrke CW, Huang LH, Ehrlich M (1982) Heat- and alkali-induced deamination of 5-methylcytosine and cytosine residues in DNA. Biochim Biophys Acta 697:371–377

Wilson VL, Jones PA (1983) Inhibition of DNA methylation by chemical carcinogens in vitro. Cell 32:239–246

Wink DA, Kasprzak KS, Maragos CM, Elespuru RK, Misra M, Dunams TM, Cebula TA, Koch WH, Andrews AW, Allen JS, Keefe LK (1991) DNA deaminating ability and genotoxicity of nitric oxide and its progenitors. Science 254:1001–1003

Wood RD, Skopek TR, Hutchison F (1984) Changes in DNA base sequence induced by targeted mutagenesis of λ phage by ultraviolet light. J Mol Biol 173:273–291

Wyszynski M, Gabbara S, Bhagwat AS (1994) Cytosine deaminations catalyzed by DNA cytosine methyltransferases are unlikely to be the major cause of mutational hot spots at sites of cytosine methylation in *Escherichia coli*. Proc Natl Acad Sci USA 91:1574–1578

Yang AS, Shen JC, Zingg JM, Mi S, Jones PA (1995) HhaI and HpaII DNA methyltransferases bind DNA mismatches, methylate uracil and block DNA repair. Nucleic Acids Res 23:1380–1387

Yang AS, Jones PA, Shibata A (1996a) The mutational burden of 5-methylcytosine. In: Russo VEA, Martienssen R, Riggs AD (eds) Epigenetic mechanisms of gene regulation. Cold Spring Harbor Laboratory Press, Cold Spring Harbor, New York, pp 77–94

Yang AS, Gonzalgo ML, Zingg JM, Millar RP, Buckley JD, Jones PA (1996b) The rate of CpG mutation in Alu repetitive elements within the p53 tumor suppressor gene in the primate germline. J Mol Biol 258:240–250

Yebra MJ, Bhagwat AS (1995) A cytosine methyltransferase converts 5-methylcytosine in DNA to thymine. Biochemistry 34:14752–14757

You YH, Halangoda A, Buettner V, Hill K, Sommer S, Pfeifer GP (1998) Methylation of CpG dinucleotides in the *lacI* gene of the Big Blue transgenic mouse. Mutation Res 420:55–65

Ziegler A, Leffell DJ, Kunala S, Sharma HW, Gailani M, Simon JA, Halperin AJ, Baden HP, Shapiro PE, Bale AE, Brash DE (1993) Mutation hot spots due to sunlight in the p53 gene of non-melanoma skin cancers. Proc Natl Acad Sci USA 90:4216–4220

Ziegler A, Jonason AS, Leffell DJ, Simon JA, Sharma HW, Kimmelman J, Remington L, Jacks T, Brash DE (1994) Sunburn and p53 in the onset of skin cancer. Nature 372:773–776

Zuo S, Boorstein RJ, Teebor GW (1995) Oxidative damage to 5-methylcytosine in DNA. Nucleic Acids Res 25:3239–3243

# The Role of DNA Methylation in Modulating Epstein-Barr Virus Gene Expression

K.D. ROBERTSON

## 1 Introduction

Epstein-Barr virus (EBV) is a ubiquitous γ herpes virus which infects over 90% of the world population (LIEBOWITZ and KIEFF 1993). The episomally maintained genome is composed of a 172-kb double-stranded DNA molecule, which was completely sequenced in 1985 and shown to code for between 80 and 100 open reading frames (BAER et al. 1984). Like other herpes viruses, EBV exists in both latent and lytic states, with only the lytic state sensitive to antiviral DNA polymerase inhibitors, such as acyclovir. EBV infection is transmitted through saliva and infects the oral epithelium and B lymphocytes via the CD21 (C3d) molecule (FINGEROTH et al. 1984; SIXBEY et al. 1984).

Infection of B cells in vitro with EBV results in outgrowth of immortalized, or continuously proliferating, lymphoblastoid cell lines that express 11 EBV transcripts. These include six nuclear antigens, [EBV nuclear antigen (EBNA)-1, 2, 3A, 3B, 3C, and leader protein], three membrane proteins [latent membrane protein (LMP)-1, 2A, and 2B], and two small, non-polyadenylated transcripts termed EBV-encoded RNA (EBER) 1 and 2. The EBNA family of proteins are driven from one of two promoters in the right end of the genome (Fig. 1A) termed the *Bam*HI C and

LME/VICHD/NIH, Bldy 18T., Rm. 106, 18 Library Dr., Bethesda, MD 20892, USA
E-mail: robertk@mail.nih.gov

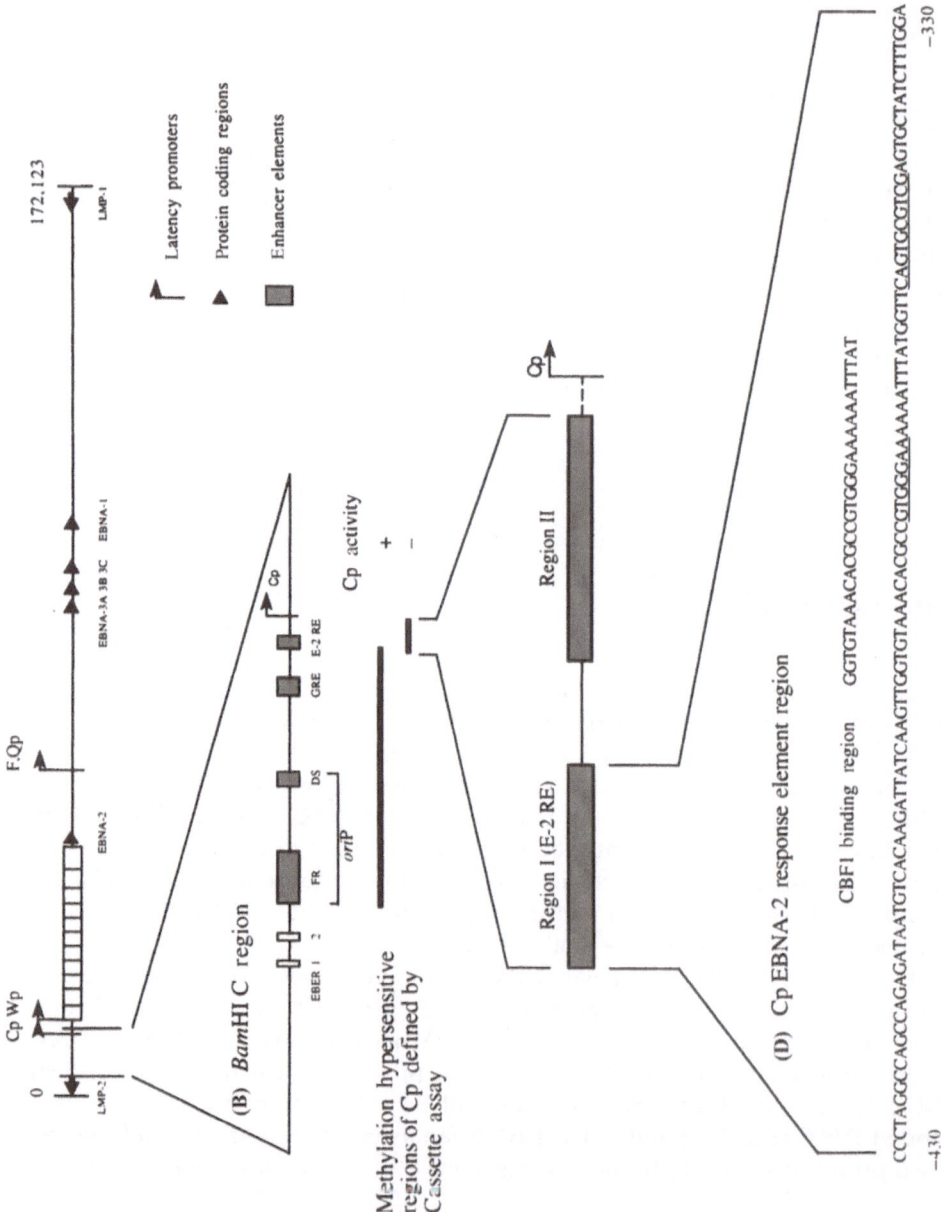

**(A) Linear EBV genome**

**(B)** *Bam*HI C region

**(C) Methylation hypersensitive regions of Cp defined by Cassette assay**

**(D) Cp EBNA-2 response element region**

CBF1 binding region    GGTGTAAACACGCCGTGGGAAAAATTTAT

CBF2 binding region    AAAAATTTATGGTTCAGTGCCGTCGAGTGCT

**CBF2 binding region    AAAAATTTATGGTTCAGTGCCGTCGAGTGCT**

**Fig. 1. A** Schematic linear representation of the Epstein-Barr-virus (EBV) genome, showing the location of most of the latency promoters and their downstream protein products *Open boxes* represent the large internal repeats. **B** Blowup of the *Bam*HI C region of the EBV genome, showing the location of known enhancer elements. FR (family of repeats) and DS (dyad symmetry element) are part of the EBV latent origin of replication *ori*P. GRE is a glucocorticoid response element, and E-2 RE is the EBV nuclear antigen 2 (EBNA-2) response element. **C** Summary of results of the methylation cassette assay (ROB-ERTSON and AMBINDER 1997a), indicating that methylation of the region encompassing the E2-RE through the C-promoter (Cp) TATA box is sufficient to shut off Cp activity (*dark bars* denote regions methylated). This region can be subdivided into two smaller regions (regions I and II). **D** Blowup of the EBNA-2 response element, showing the location of the Cp-binding factor 1 (CBF1) and CBF2 binding sites (*underlined*), as defined by electrophoretic mobility-shift oligonucleotide-competition assays (LING et al. 1993; ROBERTSON et al. 1995). Numbering is relative to the Cp TATA box at position 11,305 of the EBV genome (LIEBOWITZ and KIEFF 1993)

W promoters (Cp and Wp) because of their location in the viral *Bam*HI C and W fragments of the genome (SAMPLE et al. 1986; BODESCOT et al. 1987). Wp is used upon initial infection, then Cp is activated and remains the predominant latency promoter (WOISETSCHLAEGER et al. 1990). The EBNA family are produced by alternative splicing from a primary transcript of over 100kb (BODESCOT and PERRICAUDET 1986). It has been shown by recombinant EBV molecular genetic analysis that EBNA-1, 2, 3A, 3C, and LMP-1 are required for in vitro immortalization (MARCHINI et al. 1993; KEMPKES et al. 1995; ROBERTSON and KIEFF 1995).

Several aspects of EBV biology make it an excellent system with which to study the effects of DNA methylation on gene regulation. These include (1) knowledge of the complete sequence of the EBV genome, (2) maintenance of that genome as a mini-chromosome independent of the vast cellular genome, and (3) extensive characterization of the regulation of many important latent and lytic viral promoters by both cellular and viral factors. This chapter will highlight the past and present contributions of the EBV system to understanding of the role of DNA methylation in gene regulation and in the pathogenesis of EBV-associated malignancies.

# 2 EBV and DNA Methylation

This chapter will focus mainly on the role of DNA methylation in the regulation of the major latency *Bam*HI Cp. Results from many laboratories have shown that multiple viral and cellular factors are critical for Cp regulation. This is not surprising, given the importance of Cp-derived proteins in the viral life cycle in terms of (1) genes required for viral immortalization of B cells (EBNA-1, 2, 3A, and 3C) and (2) the potent immunogenicity of Cp-derived antigens (EBNA-2, 3A, 3B, and 3C) in the CD8(+) cytotoxic T cell (CTL) immune-control pathway in normal seropositive individuals (MURRAY et al. 1990, 1992; KHANNA et al. 1992). Further evidence that the Cp is highly regulated in vivo comes from findings indicating that no Cp derived transcripts can be detected, even by sensitive techniques like reverse-

transcriptase polymerase chain reaction (RT-PCR), in the majority of EBV-associated malignancies (African Burkitt's lymphoma, EBV–Hodgkin's disease, and
nasopharyngeal carcinoma; BROOKS et al. 1992; DEACON et al. 1993; TAO et al.
1998a). Therefore, studying the regulation of the Cp is essential for understanding
both the ability of EBV to immortalize B cells and the ability of the virus to persist
for the lifetime of the infected host in the face of a potent immune response.

Much of the original evidence for the role of DNA methylation in regulating
EBV promoters came from work at the Karolinska Institute (Sweden), and more
recent data also supports this role. For example, viral hypermethylation could be
detected in nearly all regions of the genome examined, including the BamHI C
fragment (location of the Cp) and the BamHI W fragment (location of the Wp) in
early-passage Burkitt's lymphoma cell lines, late-passage Burkitt's cell lines maintaining the highly restricted patterns of antigen expression seen in the primary
tumor (such as Rael and Mutu I), primary nasopharyngeal carcinoma tumors,
primary nasal lymphomas, and an acquired-immune-deficiency-syndrome Burkitt-
like tumor (ERNBERG et al. 1989; ALLDAY et al. 1990; MINAROVITS et al. 1994).
These studies were performed with the use of methylation-sensitive restriction enzymes followed by Southern blotting and probing with various regions of the EBV
genome. The sum of these results implicated DNA methylation as a major regulator of EBV gene expression. Similar, more recent evidence for hypermethylation
of viral sequences in EBV-associated malignancies was demonstrated for EBV-
gastric carcinoma, Hodgkin's disease, and primary Burkitt's-lymphoma tumors
(IMAI et al. 1994; ROBERTSON et al. 1996).

Massucci et al. (1989) showed that treatment of the Rael Burkitt's-lymphoma
cell line with 5-azacytidine, an inhibitor of DNA methylation, resulted in reactivation of the Cp, expression of Cp-driven antigens, and demethylation of HpaII
sites in the vicinity of Cp and other regions of the viral genome (MASUCCI et al.
1989). Antigen expression increased in a time-dependent manner, with a peak at
72h. In addition, in vitro methylation of Cp-reporter plasmids and subsequent
transfection into EBV-positive and -negative cell lines resulted in marked repression
of reporter activity. Methylation of Cp-reporter constructs in vitro with various
prokaryotic DNA methylases showed that methylation of CpG sites in the context
of the CCGG recognition site (HpaII methylase) had a more profound effect than
methylation of CpGs in the context of GCGC (HhaI methylase), indicating that the
effects of methylation on the Cp might be related to methylation of certain critical
regulatory sequences (MINAROVITS et al. 1994). The mechanism of the methylation-
mediated repression of these EBV-promoters, however, remained completely
unknown.

DNA methylation in mammals occurs at the 5 position of cytosine in the
context of CpG and is catalyzed by the DNA-methyltransferase enzyme (BIRD
1992). DNA methylation has been recognized for some time, as have its repressive
effects on transcription; however, its overall significance in gene regulation was not
universally accepted until the DNA-methyltransferase enzyme (Dnmt1) was disrupted in mice and shown to be required for embryonic development (LI et al.
1992). Approximately 70% of the CpG residues in the genome are methylated;

however, the distribution of CpG in the mammalian genome is not random, and the majority of the genome is CpG-poor or suppressed (COOPER and KRAWCZAK 1989). Certain regions of the genome have the expected CpG frequency. These regions are often clustered at the 5' ends of genes and have been termed "CpG islands" (BIRD 1986). CpG islands are not normally methylated in cells, and the mechanism preventing islands from being methylated is not fully understood (RAZIN et al. 1984; BIRD 1992; VERDINE 1994). It should be noted that EBV, unlike other human herpes viruses, is CpG suppressed over the entire genome (KARLIN et al. 1994).

# 3 DNA Methylation and the EBV Major Latency Promoter Cp

Several regulatory elements of the Cp have been characterized and include (1) the EBNA-2 response element, which will be discussed in further detail, (2) the family of EBNA-1 binding sites (family of repeats FR) localized to the latent origin of plasmid replication (*oriP*), which have been shown to transactivate the Cp in the presence of EBNA-1 (REISMAN and SUGDEN 1986; SUGDEN and WARREN 1989; SAMPLE et al. 1991), and (3) a glucocorticoid response element (GRE), which was shown to be able to enhance Cp activity in the presence of dexamethasone (KUPFER and SUMMERS 1990; Fig. 1B). Of these three major regulatory elements, only the EBNA-2 response element was shown to be affected by DNA methylation.

The minimal EBNA-2-responsive element has been defined by several groups and is approximately 100bp in length (Fig. 1D; SUNG et al. 1991; JIN and SPECK 1992). The viral transactivator EBNA-2 does not bind directly to DNA. Two cellular DNA-binding proteins were found to bind to the EBNA-2-responsive region and were termed Cp-binding factors 1 and 2 (CBF 1 and 2). Electrophoretic mobility-shift assays demonstrated that EBNA-2 interacted directly with CBF1, tethering EBNA-2 to the DNA (LING et al. 1993). Cloning of CBF1 revealed that it was identical to a previously identified protein termed recombination-signal-sequence-binding protein JK, mistakenly believed to be involved in V(D)J recombination. CBF1 is a ubiquitous cellular repressor of many genes. The binding of EBNA-2 to CBF1 masks the CBF1-repression domain while providing its own acidic transactivation domain (DOU et al. 1994; HSIEH and HAYWARD 1995). The function of CBF2, which binds almost immediately downstream of CBF1 and whose DNaseI footprint was not separable from that of CBF1 (Fig. 1D) remains unknown. However, it was demonstrated that CBF2 binding was methylation sensitive in an in vitro binding assay and that this factor was required for efficient EBNA-2-mediated Cp activity (ROBERTSON et al. 1995; FUENTES-PANANA and LING 1998). The binding of CBF1 was unaffected by methylation (ROBERTSON and AMBINDER 1997a).

Detailed mapping studies revealed that the CBF2-binding activity was sensitive to methylation of the underlined "C" within the sequence '...CAGTGCGTCG...'

(Fig. 1D) and that methylation or mutation to "T" at this position resulted in a significant reduction in EBNA-2-mediated Cp activity in a synthetic enhancer–reporter plasmid containing eight tandem copies of the Cp EBNA-2 response element driving an E1b TATA box (ROBERTSON et al. 1995). Thus, CBF2 joins a family of other transcription factors that have been shown to be methylation sensitive. These include c-Myc/Myn, AP-2, E2F, ATF/CREB-like proteins binding to cyclic-AMP responsive elements (CRE's), EBP-80, MLTF/USF, and NF-κB, major late-transcription factor/upstream stimulatory factor, and nuclear factor κB. (KOVESDI et al. 1987; WATT and MOLLOY 1988; IGUCHI-ARIGA and SCHAFFNER 1989; COMB and GOODMAN 1990; BEDNARIK et al. 1991; FALZON and KUFF 1991; PRENDERGAST et al. 1991; TATE and BIRD 1993). High-frequency methylation of the CBF2-binding site was demonstrated by bisulfite genomic sequencing (FROMMER et al. 1992) in EBV cell lines that do not express the Cp. Furthermore, the CpG site within the CBF2-binding region became demethylated at high frequency, and the Cp reactivated after treatment of cell lines with 5-azacytidine (ROBERTSON et al. 1995; SCHAEFER et al. 1997).

Is inhibition of CBF2 binding responsible for the complete inactivity of the Cp in certain EBV–Burkitt's cell lines and primary EBV malignancies? The answer would appear to be no. Other recent studies (using a technique termed the "methylation cassette assay", in which defined regions of natural Cp reporter plasmids are methylated in vitro and ligated back into their natural context and the effect upon co-transfection with and without an EBNA-2 expression vector is assessed) indicated that methylation of the EBNA-2 response element (also referred to as region I; Fig. 1C) was sufficient to abolish EBNA-2-mediated Cp activity (ROBERTSON and AMBINDER 1997a). A significant amount of basal reporter activity remained, however, and it was concluded that other regions must contribute to the complete repression of the Cp observed in vivo. Scanning of the Cp from *oriP* through the Cp TATA box indicated that a second region just upstream of the TATA box, in combination with the repression associated with the EBNA-2 response element, was sufficient to account for nearly 100% of the methylation-mediated repression of the Cp (ROBERTSON and AMBINDER 1997a). A summary of the important regions of the Cp and the methylation-hypersensitive regions is shown in Fig. 1B,C. The exact mechanism of repression associated with methylation of the downstream region (also referred to as region II; Fig. 1C) appeared to involve the binding of a repressor protein specifically to methylated DNA (ROBERTSON and AMBINDER 1997a; Robertson and Ambinder, unpublished) in a manner that may be similar to the methyl-binding proteins MeCP1 and MeCP2, which have been shown to act as transcriptional repressors (BOYES and BIRD 1991; MEEHAN et al. 1992). Methylation of the FR region of *oriP* also had no significant repressive effects on Cp activity. Other systems investigated in a similar manner have indicated that methylation may act primarily by a non-specific mechanism, perhaps by recruiting non-sequence-specific methyl-binding proteins like MeCP1 or by formation of inactive heterochromatin, which is incompatible with transcription (BIRD 1992, 1993; KASS et al. 1993). Studies performed with the methylation cassette assay, the most detailed study of any mammalian cellular or viral promoter so far, indicated that methylation of very discrete regions of the promoter could account for all of the methylation-mediated repression observed (ROBERTSON and AMBINDER 1997a). Clearly, other factors, including non-specific ones, may be involved in the repression of the Cp in vivo in tumors, but the vast majority of the repressive effects appear to be mediated by methylation of highly defined regions within the promoter.

It should also be noted here that methylation of other EBV promoters, including the latency promoters Wp (MASUCCI et al. 1989; JANSSON et al. 1992), and Qp (Fig. 1A; TAO et al. 1998b), the enhancer/promoter regions associated with the lytic EBV origin of replication (*oriLyt*, not shown; NONKWELO and LONG 1993), and the promoter for the EBV transforming protein LMP-1 (Fig. 1A; FALK et al. 1998), has been associated with transcriptional silence. Modulation of these and other EBV promoters by DNA methylation may play equally important roles in the EBV life cycle. For example, methylation of the lytic origin of replication may regulate the latent–lytic life cycle choice; however, these systems are not as well characterized as the Cp and will not be discussed here due to space constraints.

# 4 Methylation of the Cp in EBV-Associated Tumors

The correlation between Cp expression and Cp methylation at the CBF2-binding site and surrounding sites was nearly perfect in EBV cell lines (ROBERTSON et al. 1995); however, extrapolation of patterns of methylation from cell lines to tumor tissue can be hazardous. Adaptation to cell culture is generally associated with methylation of CpG islands. This increase in methylation of cellular CpG islands with passage in tissue cultures has been extensively documented in murine and human tissue-culture systems including B lymphocytes immortalized in vitro with EBV (ANTEQUERA et al. 1990; JONES et al. 1990). For this reason, extensive genomic sequencing of the Cp region, including the CBF2-binding site, was carried out on primary EBV tumors. As shown in Table 1, in a 253-bp region of the Cp EBNA-2 response element containing nine CpG sites, genomic sequencing revealed that, in Burkitt's lymphoma, Hodgkin's disease, EBV-lymphoproliferative disease (LPD), and nasopharyngeal carcinoma tumor samples, the majority of these CpG sites were methylated. Furthermore, the CpG site implicated in CBF2 binding (site 4 in Table 1) was methylated in an even higher percentage of clones (approaching 90%). That methylation patterns can differ between tumor types was evidenced by results for a CpG site upstream of site 4, referred to as site 2. This site was only sporadically methylated in Burkitt's lymphoma but was heavily methylated in Hodgkin's disease and nasopharyngeal carcinoma (Table 1). In contrast, LPD-tumor samples demonstrating Cp activity by RT-PCR were almost completely devoid of methylation at all sites, particularly at site 4 (ROBERTSON et al. 1996; Robertson and Ambinder, unpublished).

**Table 1.** Summary of genomic sequencing data in the C promoter EBNA-2 response-element region from primary EBV tumor samples

| Tumor type | % Methylated clones | | | |
| --- | --- | --- | --- | --- |
| | Total CpG sites examined (9) | CpG site 2 | CpG site 4 (CBF2) | % Of clones devoid of methylation |
| Burkitt's | 68.3 | 18.5 | 81.5 | 7.4 |
| Hodgkin's | 82.6 | 79.5 | 89.7 | 10.3 |
| LPD (Cp+) | 22.2 | 0.0 | 0.0 | 44.4 |
| LPD (Cp−) | 87.0 | 16.7 | 100.0 | 0.0 |
| NPC | 72.2 | 73.5 | 67.6 | 23.5 |
| Normal PBMC | 48.0 | 29.4 | 50.0 | 45.5 |

*CBF2*, Cp-binding factor 2; *LPD*, EBV lymphoproliferative disease displaying Cp activity (Cp+) or lacking Cp activity (Cp−), as defined by reverse-transcriptase polymerase chain reaction; *NPC*, nasopharyngeal carcinoma; *PBMC*, peripheral blood mononuclear cells from normal seropositive individuals.

# 5 Methylation of the Cp in Normal Seropositive Individuals and a Model for EBV Persistence

Since primary infection of B lymphocytes in vitro leads to the establishment of EBV-infected B lymphoblastoid cell lines in which the viral genome is not methylated, there has been a presumption that methylation of the viral genome was a phenomenon confined to tumors or tumor cell lines. Aberrant methylation of cellular genes and their transcriptional-regulatory regions is a well-recognized phenomenon in tumors and tumor cell lines (ROBERTSON and JONES 1997). Evidence has recently emerged suggesting that methylation of transcriptional-regulatory regions of tumor-suppressor genes may be a step in multi-step tumorigenesis akin to deletion or mutation of tumor-suppressor genes (MYOHANEN et al. 1998). Methylation of viral promoters suppressing transcription of immunodominant antigens might be another way in which aberrant methylation in tumors might contribute to tumorigenesis. This issue was addressed by bisulfite genomic sequencing of EBV DNA extracted from the peripheral blood mononuclear cells (PBMCs) of healthy, seropositive individuals. Analysis of the methylation status of the Cp in PBMCs from healthy, seropositive adults (Table 1) showed 48% methylation when the nine CpG sites in the region examined were taken together. The CpG site important for the binding of CBF2 was methylated in 50% of the cloned inserts. Unlike the situation in tumors, almost half (45%) of the cloned inserts from peripheral blood demonstrated a complete absence of methylation (Table 1). However, if at least one site was methylated, then the remainder of possible sites were likely to be methylated also (89%; ROBERTSON and AMBINDER 1997b). Thus, methylation of the EBV genome was not restricted to tumors but appeared to be an integral part of the viral lifecycle. This is in keeping with an analysis of CpG suppression of DNA viruses, which showed that EBV was the only human herpes virus demonstrating CpG suppression (KARLIN et al. 1994). Methylation over evolutionary time is believed to result in suppression of the CpG-dinucleotide sequence in vertebrates and is generally believed to be a consequence of methylation at position five of the cytosine base, which following deamination and failure to repair the mismatch becomes TpG/CpA (SHEN et al. 1992, 1994; JIRICNY 1996). No other large human virus shows this CpG suppression, and the other large primate DNA viruses which are CpG suppressed are also γ herpes viruses that have in common with EBV latency in lymphocytes (KARLIN et al. 1994). These results taken together, support the notion that methylation, of at least the Cp region of the viral genome, is an integral part of the viral life cycle.

These observations led to the following model (Fig. 2). It is proposed that there exists a reservoir of latently infected B lymphocytes in which the C promoter is methylated and silent (veiled latency). The infected cells are thus insulated from CD8(+) cytotoxic T cell immune surveillance. In response to some stimulus or

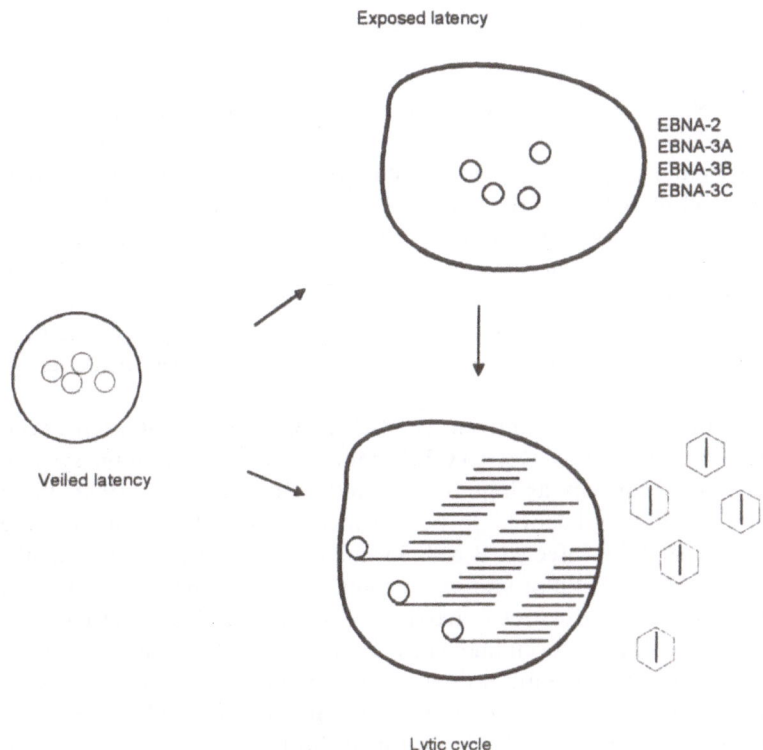

**Fig. 2.** Model for Epstein-Barr-virus (EBV) persistence in peripheral blood lymphocytes of healthy, seropositive individuals (adapted from ROBERTSON and AMBINDER 1997b). Lymphocytes infected with EBV are suggested to exist in three different states. In veiled latency, the immunodominant EBV nuclear antigens (EBNAs) are not expressed, the C promoter is methylated, and its silence enforced. In exposed latency, the immunodominant EBNAs are expressed and the C promoter is not methylated. In the third state (lytic replication with the production of new virions), it appears that the EBV genome is not methylated (Robertson and Ambinder, unpublished data)

perhaps by a random process, viral episomes in these cells demethylate, leading either to Cp activation with expression of latency genes that render the cells sensitive to immune surveillance but which also drive proliferation of the infected cells (exposed latency) or to lytic activation, with production of new virions (ROBERTSON and AMBINDER 1997b). The balance between latent and lytic reactivation may be determined by local environmental interactions. This model accommodates an equilibrium between methylated Cp latency, in which the immortalizing viral genes are not expressed, and an unmethylated latency, in which the immortalizing viral genes are expressed but the infected B cells are presumably very short lived because of cytotoxic T cell immune surveillance (MURRAY et al. 1990; KHANNA et al. 1992; MURRAY et al. 1992). There is also evidence that the viral DNA is hypomethylated during lytic infection (Robertson and Ambinder, unpublished).

# 6 Therapeutic Implications for EBV-Associated Malignancies

Inhibitors of the DNA methyltransferase have been known for some time. These include the nucleotide analogs 5-azacytidine and 5-aza-2'-deoxycytidine, which act as suicide substrates and trap the enzyme in a covalent complex on DNA (SMITH et al. 1992). This effect requires that the analogs be incorporated into the DNA and results in a progressive loss of methylation following each round of cell division due to the lack of re-methylation of newly synthesized daughter strands (JONES 1985). Both analogs exert their effects in a similar manner by virtue of the substitution of nitrogen for carbon at the 5-position; however, 5-azacytidine is less potent than the deoxy version because it must first be converted by cellular enzymes to the deoxy form (RAZIN et al. 1984; CHABNER 1990).

The immunogenic, Cp-derived antigens (EBNA-2, 3A, 3B, and 3C) have been shown to be potent targets of the CD8( + ) CTL arm of the cellular immune system, with a CTL precursor frequency as high as 1 in 400 (MURRAY et al. 1990, 1992; KHANNA et al. 1992). This is believed to be the mechanism which counters the proliferative capacity of EBV-infected B cells in normal, seropositive individuals. The combination of a strong cellular immune response directed against Cp-derived antigens and the ability to modulate Cp activity by the use of DNA-methyltransferase inhibitors in EBV–Burkitt's cell lines (MASUCCI et al. 1989; ROBERTSON et al. 1995) provides a unique therapeutic strategy. Treatment of individuals having EBV-associated malignancies with 5-aza-2'-deoxycytidine could result in reactivation of immunogenic Cp-derived antigens, restoring both immune recognition of the tumor and killing by CD8( + ) T cells. This killing would be highly specific for the EBV-infected tumor cells. An extension of this idea comes from an observation, made in two laboratories, that treating Burkitt's cell lines maintaining the phenotype of the original tumor with 5-aza-2'-deoxycytidine results in induction of the viral lytic cycle (MASUCCI et al. 1989 and Robertson and Ambinder unpublished data). This could further enhance tumor cell killing by (1) viral-induced lysis of tumor cells, (2) increased levels of CTL target antigens, i.e., the lytic transcripts (BOGEDAIN et al. 1995), and (3) enhanced killing by additional treatment with chain-terminating nucleoside analogs, such as gancyclovir, which are specifically phosphorylated by the lytic-cycle thymidine kinase. This form of therapy would seem to be applicable to nearly all EBV-associated malignancies because, with the exception of some EBV lymphomas in the immunocompromised, the Cp is inactive and hypermethylated. A recent in vitro study using EBV cell lines supports such a gancyclovir-based approach to enhancing tumor cell killing by induction of lytic infection (WESTPHAL et al. 1999). Although bone-marrow toxicity has been observed with these agents when used for the treatment of certain leukemias, both methyltransferase inhibitors have been used extensively in clinical trials of therapies ranging from anticancer therapy to demethylation and re-expression of embryonic γ-globin in the treatment of hemoglobinopathies (Chap. 10; LEY et al. 1982; HUMPHRIES et al. 1985; PINTO and ZAGONEL 1993).

*Acknowledgements.* I would like to thank Richard Ambinder and Qian Tao for helpful comments in the preparation of this chapter. KDR was supported by an American Cancer Society postdoctoral fellowship.

# References

Allday MJ, Kundu D, Finerty S, Griffin BE (1990) CpG methylation of viral DNA in EBV-associated tumors. Int J Cancer 45:1125–1130

Antequera F, Boyes J, Bird A (1990) High levels of de novo methylation and altered chromatin structure at CpG islands in cell lines. Cell 62:503–514

Baer R, Bankier AT, Biggin MD, Deininger PL, Farrell PJ, Gibson TJ, Hatfull G, Hudson GS, Satchwell SC, Seguin C, Tuffnell PS, Barrell BG (1984) DNA sequence and expression of the B95–8 Epstein-Barr-virus genome. Nature 310:207–211

Bednarik DP, Duckett C, Kim SU, Perez VL, Griffis K, Guenther PC, Folks TM (1991) DNA CpG methylation inhibits binding of NF-κB proteins to the HIV-1 long terminal repeat cognate DNA motifs. N Biol 3:969–976

Bird A (1986) CpG-rich islands and the function of DNA methylation. Nature 321:209–213

Bird A (1992) The essentials of DNA methylation. Cell 70:5–8

Bird AP (1993) Functions for DNA methylation in vertebrates. Cold Spring Harbor symposium on quantitative biology LVIII:281–285

Bodescot M, Perricaudet M (1986) Epstein-Barr-virus mRNAs produced by alternative splicing. Nucleic Acids Res 14:7103–7115

Bodescot M, Perricaudet M, Farrell PJ (1987) A promoter for the highly spliced EBNA family of RNAs of Epstein-Barr virus. J Virol 61:3424–3430

Bogedain C, Wolf H, Modrow S, Stuber G, Jilg W (1995) Specific cytotoxic T lymphocytes recognize the immediate-early transactivator Zta of Epstein-Barr virus. J Virol 69:4872–4879

Boyes J, Bird A (1991) DNA methylation inhibits transcription indirectly via a methyl-CpG binding protein. Cell 64:1123–1134

Brooks L, Yao QY, Rickinson AB, Young LS (1992) Epstein-Barr virus latent gene transcription in nasopharyngeal carcinoma cells: coexpression of EBNA-1, LMP1, and LMP2 transcripts. J Virol 66:2689–2697

Chabner BA (1990) Cytidine analogs. In: Chabner BA, Collins JM (eds) Cancer chemotherapy: principles and practice. Lippincott, Philadelphia

Comb M, Goodman HM (1990) CpG methylation inhibits proenkephalin gene expression and binding of the transcription factor AP-2. Nucleic Acids Res 18:3975–3982

Cooper DN, Krawczak M (1989) Cytosine methylation and the fate of CpG dinucleotides in vertebrate genomes. Hum Genet 83:181–188

Deacon EM, Pallesen G, Niedobitek G, Crocker J, Brooks L, Rickinson AB, Young LS (1993) Epstein-Barr virus and Hodgkin's disease: transcriptional analysis of virus latency in the malignant cells. J Exp Med 177:339–349

Dou S, Zeng X, Cortes P, Erdjument-Bromage H, Tempst P, Honjo T, Vales LD (1994) The recombination signal sequence-binding protein RBP-2N functions as a transcriptional repressor. Mol Cell Biol 14:3310–3319

Ernberg I, Falk K, Minarovits J, Busson P, Tursz T, Masucci MG, Klein G (1989) The role of methylation in the phenotype-dependent modulation of Epstein-Barr nuclear-antigen-2 and latent-membrane-protein genes in cells latently infected with Epstein-Barr virus. J Gen Virol 70:2989–3002

Falk KI, Szekely L, Aleman A, Ernberg I (1998) Specific methylation patterns in two control regions of Epstein-Barr virus: the LMP-1-coding upstream regulatory region and an origin of DNA replication (oriP). J Virol 72:2969–2974

Falzon M, Kuff EL (1991) Binding of the transcription factor EBP-80 mediates the methylation response of an intracisternal A-particle long terminal repeat promoter. Mol Cell Biol 11:117–125

Fingeroth JD, Weis JJ, Tedder TF, Strominger JL, Biro PA, Fearon DT (1984) Epstein-Barr-virus receptor of human B-lymphocytes is the C3d receptor CR2. Proc Natl Acad Sci USA 81: 4510–4514

Frommer M, McDonald LE, Millar DS, Collis CM, Watt F, Grigg GW, Molloy PL, Paul CL (1992) A genomic sequencing protocol that yields a positive display of 5-methylcytosine residues in individual DNA strands. Proc Natl Acad Sci USA 89:1827–1831

Fuentes-Panana EM, Ling PD (1998) Characterization of the CBF2-binding site within the Epstein-Barr-virus latency C promoter and its role in modulating EBNA 2-mediated transactivation. J Virol 72:693–700

Hsieh JD, Hayward SD (1995) Masking of the CBF1/RBPJ$_K$ transcriptional repression domain by Epstein-Barr virus EBNA2. Science 268:560–563

Humphries RK, Dover G, Young NS, Moore JG, Charache S, Ley T, Neinhuis AW (1985) 5-azacytidine acts directly on both erythroid precursors and progenitors to increase production of fetal hemoglobin. J Clin Invest 75:547–557

Iguchi-Ariga S, Schaffner W (1989) CpG methylation of the cAMP-responsive enhancer/promoter sequence TGACGTCA abolishes specific factor binding as well as transcriptional activation. Genes Dev 3:612–619

Imai S, Koizumi S, Sugiura M, Tokunaga M, Uemura Y, Yamamoto N, Tanaka S, Sato E, Osato T (1994) Gastric carcinoma: Monoclonal epithelial malignant cells expressing Epstein-Barr-virus latent infection protein. Proc Natl Acad Sci USA 91:9131–9135

Jansson A, Masucci M, Rymo L (1992) Methylation of discrete sites within the enhancer region regulates the activity of the Epstein-Barr virus *BamHI* W promoter in Burkitt lymphoma lines. J Virol 66:62–69

Jin XW, Speck SH (1992) Identification of critical *cis* elements involved in mediating Epstein-Barr virus nuclear antigen 2-dependent activity of an enhancer located upstream of the viral *BamHI* C promoter. J Virol 66:2846–2852

Jiricny J (1996) Mismatch repair and cancer. In: Cancer surveys: genetic instability in cancer. Imperial Cancer Research Fund

Jones PA (1985) Altering gene expression with 5-azacytidine. Cell 40:485–486

Jones PA, Wolkowicz MJ, Rideout WM, Gonzales FA, Marziasz CM, Coetzee G, Tapscott SJ (1990) De novo methylation of the *MyoD1* CpG island during the establishment of immortal cell lines. Proc Natl Acad Sci USA 87:6117–6121

Karlin S, Doerfler W, Cardon LR (1994) Why is CpG suppressed in the genomes of virtually all small eukaryotic viruses but not those of large eukaryotic viruses. J Virol 68:2889–2897

Kass SU, Goddard JP, Adams RLP (1993) Inactive chromatin spreads from a focus of methylation. Mol Cell Biol 13:7372–7379

Kempkes B, Pich D, Zeidler R, Sugden B, Hammerschmidt W (1995) Immortalization of human B lymphocytes by a plasmid containing 71 kilobase pairs of Epstein-Barr virus DNA. J Virol 69: 231–238

Khanna R, Burrows SR, Kurilla MG, Jacob CA, Misko IS, Sculley TB, Kieff E, Moss DJ (1992) Localization of Epstein-Barr virus cytotoxic T cell epitopes using recombinant vaccinia: implications for vaccine development. J Exp Med 176:169–176

Kovesdi I, Reichel R, Nevins JR (1987) Role of an adenovirus E2-promoter-binding factor in E1A-mediated coordinate gene control. Proc Natl Acad Sci USA 84:2180–2184

Kupfer SR, Summers WC (1990) Identification of a glucocorticoid-responsive element in Epstein-Barr virus. J Virol 64:1984–1990

Ley TJ, DeSimone J, Anagnou NP, Keller GH, Humphries K, Turner PH, Young NS, Heller PH, Neinhuis AW (1982) 5-azacytidine selectively increases γ-globin synthesis in a patient with β⁺ thalassemia. N Engl J Med 307:1469–1475

Li E, Bestor TH, Jaenisch R (1992) Targeted mutation of the DNA methyltransferase gene results in embryonic lethality. Cell 69:915–926

Liebowitz D, Kieff E (1993) Epstein-Barr virus. In: Roizman B, Whitley RJ, Lopez C (eds) The human herpes viruses. Raven Press Ltd., New York

Ling PD, Rawlins DR, Hayward SD (1993) The EBV immortalizing protein EBNA-2 is targeted to DNA by a cellular enhancer-binding protein. Proc Natl Acad Sci USA 90:9237–9241

Marchini A, Kieff E, Longnecker R (1993) Marker rescue of a transformation-negative Epstein-Barr virus recombinant from an infected Burkitt lymphoma cell line: a method useful for analysis of genes essential for transformation. J Virol 67:606–609

Masucci MG, Contreras-Salazar B, Ragner E, Falk K, Minarovits J, Ernberg I, Klein G (1989) 5-Azacytidine up regulates the expression of Epstein-Barr virus nuclear antigen 2 (EBNA-2) through EBNA-6 and latent membrane protein in the Burkitt's lymphoma line Rael. J Virol 63:3135–3141

Meehan RR, Lewis JD, Bird AP (1992) Characterization of MeCP2, a vertebrate DNA binding protein with affinity for methylated DNA. Nucleic Acids Res 20:5085–5092

Minarovits J, Hu L, Minarovits-Kormuta S, Klein G, Ernberg I (1994a) Sequence-specific methylation inhibits the activity of the Epstein-Barr virus LMP 1 and BCR2 enhancer-promoter regions. Virology 200:661–667

Minarovits J, Hu L-F, Imai S, Harabuchi Y, Kataura A, Minarovits-Kormuta S, Osato T, Klein G (1994b) Clonality, expression and methylation patterns of the Epstein-Barr-virus genomes in lethal midline granulomas classified as peripheral angiocentric T cell lymphomas. J Gen Virol 75:77–84

Murray RJ, Kurilla MG, Griffin HM, Brooks JM, Mackett M, Arrand JR, Rowe M, Burrows SR, Moss DJ, Kieff E, Rickinson AB (1990) Human cytotoxic T-cell response against Epstein-Barr virus nuclear antigens demonstrated by using recombinant vaccinia viruses. Proc Natl Acad Sci USA 87:2906–2910

Murray RJ, Kurilla MG, Brooks JM, Thomas WA, Rowe M, Kieff E, Rickinson AB (1992) Identification of target antigens for the human cytotoxic T cell response to Epstein-Barr virus (EBV): implications for the immune control of EBV-positive malignancies. J Exp Med 176:157–168

Myohanen SK, Baylin SB, Herman JG (1998) Hypermethylation can selectively silence individual p16^{INK4a} alleles in neoplasia. Cancer Res 58:591–593

Nonkwelo CB, Long WK (1993) Regulation of Epstein-Barr-virus BamHI-H divergent promoter by DNA methylation. Virology 197:205–215

Pinto A, Zagonel V (1993) 5-Aza-2′-deoxycytidine (Decitabine) and 5-azacytidine in the treatment of acute myeloid leukemias and myelodysplastic syndromes: past, present and future trends. Leukemia 7:51–60

Prendergast GC, Lawe D, Ziff EB (1991) Association of Myn, the murine homolog of Max, with c-Myc stimulates methylation-sensitive DNA binding and ras cotransformation. Cell 65:395–407

Razin A, Cedar H, Riggs AD (1984) Gene activation by 5-azacytidine. In: Jones PA (ed) DNA methylation: biochemistry and biological significance. Springer-Verlag, New York

Reisman D, Sugden B (1986) trans Activation of an Epstein-Barr-viral transcriptional enhancer by the Epstein-Barr-viral nuclear antigen 1. Mol Cell Biol 6:3838–3846

Robertson ES, Kieff ED (1995) Genetic analysis of Epstein-Barr virus in B lymphocytes. Epstein-Barr Virus Rep 2:73–80

Robertson KD, Ambinder RF (1997a) Mapping promoter regions that are hypersensitive to methylation-mediated inhibition of transcription: application of the methylation cassette assay to the Epstein-Barr-virus major latency promoter. J Virol 71:6445–6454

Robertson KD, Ambinder RF (1997b) Methylation of the Epstein-Barr virus genome in normal lymphocytes. Blood 90:4480–4484

Robertson KD, Jones PA (1997) Dynamic interrelationships between DNA replication, methylation, and repair. Am J Hum Genet 61:1220–1224

Robertson KD, Hayward SD, Ling PD, Samid D, Ambinder RF (1995) Transcriptional activation of the Epstein-Barr-virus latency C promoter after 5-azacytidine treatment: evidence that demethylation at a single CpG site is crucial. Mol Cell Biol 15:6150–6159

Robertson KD, Manns A, Swinnen LJ, Zong JC, Gulley ML, Ambinder RF (1996) CpG methylation of the major Epstein-Barr-virus latency promoter in Burkitt's lymphoma and Hodgkin's disease. Blood 88:3129–3136

Sample J, Hummel M, Braun D, Birkenbach M, Kieff E (1986) Nucleotide sequences of mRNAs encoding Epstein-Barr-virus nuclear proteins: a probable transcriptional-initiation site. Proc Natl Acad Sci USA 83:5096–5100

Sample J, Brooks L, Sample C, Young L, Rowe M, Gregory C, Rickinson A, Kieff E (1991) Restricted Epstein-Barr-virus protein expression in Burkitt lymphoma is due to a different Epstein-Barr nuclear antigen 1 transcriptional-initiation site. Proc Natl Acad Sci USA 88:6343–6347

Schaefer BC, Strominger JL, Speck SH (1997) Host-cell-determined methylation of specific Epstein-Barr-virus promoters regulates the choice between distinct viral latency programs. Mol Cell Biol 17: 364–377

Shen J-C, Rideout WM, Jones PA (1992) High-frequency mutagenesis by a DNA methyltransferase. Cell 71:1073–1080

Shen J-C, Rideout WM, Jones PA (1994) The rate of hydrolytic deamination of 5-methylcytosine in double-stranded DNA. Nucleic Acids Res 22:972–976

Sixbey JW, Nedrud JG, Raab-Traub N, Hanes RA, Pagano JS (1984) Epstein-Barr-virus replication in oropharyngeal epithelial cells. N Engl J Med 310:1225–1230

Smith SS, Kaplan BE, Sowers LC, Newman EM (1992) Mechanism of human methyl-directed DNA methyltransferase and the fidelity of cytosine methylation. Proc Natl Acad Sci USA 89:4744–4748

Sugden B, Warren N (1989) A promoter of Epstein-Barr virus that can function during latent infection can be transactivated by EBNA-1, a viral protein required for viral DNA replication during latent infection. J Virol 63:2644–2649

Sung NS, Kenney S, Gutsch D, Pagano JS (1991) EBNA-2 transactivates a lymphoid-specific enhancer in the *BamHI* C promoter of Epstein-Barr virus. J Virol 65:2164–2169

Tao Q, Robertson KD, Manns A, Hildesheim A, Ambinder RF (1998a) Epstein Barr virus (EBV) in endemic Burkitt's lymphoma: molecular analysis of primary tumor tissue. Blood 91:1737–1381

Tao Q, Robertson KD, Manns A, Hildesheim A, Ambinder RF (1998b) The Epstein-Barr virus (EBV) latent promoter Qp is constitutively active, hypomethylated, and methylation sensitive. J Virol 72:7075–7083

Tate PH, Bird AP (1993) Effects of DNA methylation on DNA-binding proteins and gene expression. Curr Opin Genet Dev 3:226–231

Verdine GL (1994) The flip side of DNA methylation. Cell 76:197–200

Watt F, Molloy PL (1988) Cytosine methylation prevents binding to DNA of a HeLa cell transcription factor required for optimal expression of the adenovirus major late promoter. Genes Dev 2:1136–1143

Westphal E-M, Mauser A, Swenson J, Davis MG, Talarico CL, Kenney SC (1999) Induction of lytic Epstein-Barr-virus (EBV) infection in EBV-associated malignancies using adenovirus vector in vitro and in vivo. Canc Res 59:1485–1491

Woisetschlaeger M, Yandava CN, Furmanski LA, Strominger JL, Speck SH (1990) Promoter switching in Epstein-Barr virus during the initial stages of infection of B lymphocytes. Proc Natl Acad Sci USA 87:1725–1729

# Promoter-Region Hypermethylation and Gene Silencing in Human Cancer

J.G. HERMAN and S.B. BAYLIN

# 1 Introduction

As has been outlined in recent reviews (JONES 1996; BAYLIN et al. 1998), changes in the status of DNA methylation are one of the most common molecular alterations in human neoplasia. While, as discussed elsewhere in this book, a prominent aspect of this change involves widespread loss of methyl groups from the cancer cell genome, other more localized genomic regions simultaneously undergo increases in methylation, and these are the subject of the current chapter. This hypermethylation, which is being increasingly recognized in neoplastic cells, affects gene-promoter regions and is being revealed as one of the most frequent mechanisms of loss of gene function (including inactivation of tumor-suppressor genes) in cancer. In the present chapter, we will define this pattern of hypermethylation, outline the genes involved, and discuss what is known about the mechanisms through which

The Johns Hopkins Oncology Center, 424 N. Bond Street, Baltimore, MD 21231, USA this alteration is associated with transcriptional silencing of genes and the processes that may underlie this DNA change during tumor progression. We will particularly stress the emerging evidence that this epigenetic change works closely with genetic alterations to drive tumorigenesis. We will also discuss the great promise that DNA hypermethylation holds for providing very sensitive molecular markers for early tumor detection, monitoring prognosis, and assessing the efficacy of prevention strategies. Finally, we will consider the possibility that this DNA modification and the chromatin changes associated with it can be targeted for therapeutic purposes.

## 2  Hypermethylation of CpG Islands in Neoplastic Cells and the Genes Involved

As outlined elsewhere in this book, cytosines are methylated in the human genome only when located 5' to a guanosine and when the CpG nucleotide has been severely depleted (through evolution) in the vertebrate genome to less than 10% of the predicted frequency. The majority of the remaining CpG dinucleotides (over 70%) are heavily methylated throughout most of the human genome (BIRD 1987, 1992). However, in small stretches of DNA termed CpG islands, usually from 500bp to 2000bp in length, the CpG dinucleotide occurs at an expected or increased frequency, and these areas are frequently located in and around the transcription-start site of approximately half of (or some 40,000) human genes (BIRD 1987, 1992). For virtually all of these genes (except for non-transcribed genes on the inactive X-chromosome of the female and the transcriptionally silenced allele for selected autosomal genes that are imprinted), these promoter-region CpG islands are maintained free of methylation in normal cells of all types. This is the case regardless of whether these genes are transcribed (BIRD 1987, 1992). This unmethylated status is thought to be a prerequisite state in order for the involved genes to be maintained in an actively transcribed or transcription-ready state (BIRD 1987, 1992; ANTEQUERA and BIRD 1993). Methylation within promoter-region CpG islands, especially when dense, is usually correlated with downregulated or silenced transcription, as noted for the X-chromosome and imprinted genes mentioned above. It should be noted that, in the 50% of genes that have a depleted content of CpG sites in the promoter region, these cytosines can frequently be differentially methylated contrary to the expression status of the gene in normal tissues (CEDAR 1988; KAFRI et al. 1993). The mechanisms through which methylation in promoter CpG islands or other promoter CpG sites participates in loss of transcription capacity will be discussed in a later section.

Over the past 3–4 years, it has been increasingly recognized that the CpG islands of a growing number of genes, which are unmethylated in normal tissues, are methylated to varying degrees in multiple types of human cancer (JONES 1996; BAYLIN et al. 1998; JONES and LAIRD 1999). Where examined in detail, this apparently aberrant methylation is often associated with a markedly reduced and/or completely silenced expression of the involved gene in the tumor cells as compared with the normal tissue from which the tumor arose (JONES 1996; BAYLIN et al. 1998; JONES and LAIRD 1999). It is not clear that loss of expression of every such hypermethylated gene is fundamental to formation of the tumor type studied. However, there are now excellent examples of critical tumor-suppressor genes (Table 1), each associated with specific genetic forms of cancer when mutated in the germline of involved kindreds, where hypermethylation and transcriptional silencing is associated with loss of gene function in sporadic forms of the same or additional kinds of human tumors. In the sporadic tumors, the hypermethylation of one allele is often accompanied by deletion of the opposite allele (HERMAN et al. 1994; MERLO et al. 1995), mimicking the type of loss of heterozygosity usually seen with loss of tumor-suppressor gene-function due to genetic alterations. To date, the cancer-associated genes most frequently associated with hypermethylation and loss of function in human neoplasia are the *p16* gene (GONZALEZ-ZULUETA et al. 1995; HERMAN et al. 1995, 1997; MERLO et al. 1995; BAYLIN et al. 1998), which encodes a cyclin-dependent kinase inhibitor (SERRANO et al. 1993, 1996; STRAUSS et al. 1995; WEINBERG 1995), the E-cadherin gene (GRAFF et al. 1995, 1998; YOSHIURA et al. 1995), which encodes a cell-surface adhesion protein that may protect against the invasive behavior of cancer cells (VLEMINCKX et al. 1991; MAREEL et al. 1995), and the *hMLH1* gene (KANE et al. 1997; HERMAN et al. 1998), which encodes a key mismatch-repair gene (MODRICH and LAHUE 1996; THOMAS et al. 1996).

The importance of promoter methylation as a mechanism associated with gene inactivation in cancer is emphasized by three important characteristics of this process as it effects key tumor-suppressor genes. First, the gene silencing associated with the promoter hypermethylation is a fully heritable event for individual alleles of a given gene. This is best evidenced by our observation that, in a long-established colon cancer cell line, a mutated copy of the *p16*-gene allele has an unmethylated promoter CpG island and is expressed, while the wild-type gene is hypermethylated and silenced (MYOHANEN et al. 1998). Second, the hypermethylation change virtually always occurs only in those tumors which lack coding-region mutations in the involved gene (BAYLIN et al. 1998). Third, the selective advantage for loss of

**Table 1.** Known cancer-causing genes proven to be silenced in association with promoter hypermethylation in sporadic tumor types. In each of these cases, the silencing of the gene, reactivation with demethylation, inverse relationship to the presence of mutations, and selective advantage of the promoter hypermethylation silencing have been shown for the sporadic tumors involved. References for *p16^{INK4a}*, *E-cadherin*, *VHL*, and *hMLH1* are found in the text

| Gene | Familial syndrome with germline mutation | Sporadic tumors with hypermethylation |
|------|------------------------------------------|---------------------------------------|
| *p16^{INK4a}* | Melanoma | Virtually all types – varying percentages (Fig. 1) |
| *E-cadherin* | Diffuse gastric carcinoma | Most epithelial tumor types – varying percentages |
| *VHL* | Von-Hippel Lindau (clear cell renal carcinoma) | Clear cell renal carcinoma (5–20%) |
| *hMLH1* | HNPCC – MI+ colon cancer | MI+ colon (>70%), endometrial (>90%), and gastric (>80%) carcinomas |

*HNPCC*, hereditary non-polyposis colorectal cancer; *MI*, microsatellite instability.

gene function associated with the methylation change appears to be identical to loss of function via coding-region mutations (BAYLIN et al. 1998). Examples include the facts that:

1. Both coding-region mutations and hypermethylation of the *VHL* gene are virtually specific for clear cell renal cancer (HERMAN et al. 1994).
2. Both mutations and hypermethylation of the cyclin-dependent kinase-inhibitor-encoding gene, *p16*, occur only in the absence of mutations in the *Rb* gene (HERMAN et al. 1995; BAYLIN et al. 1998). *Rb* and *p16* both encode for major control proteins in the cyclin-D–Rb cell-cycle-control pathway, and tumors need to lose function for only one of these genes in order to deregulate cycle control (KAMB et al. 1994; SHERR 1994; STRAUSS et al. 1995).
3. Both hypermethylation and genetic changes in the mismatch repair gene *hMLH1* occur frequently only in colon, gastric and endometrial carcinomas, the only cancers which have a high incidence of microsatellite instability resulting from loss of this repair-gene function (CUNNINGHAM et al. 1998; ESTELLER et al. 1998a; HERMAN et al. 1998; FLEISHER et al. 1999; LEUNG et al. 1999; SIMPKINS et al. 1999).

For the major forms of human cancer, as shown in Fig. 1, aberrant promoter-region methylation must be considered in addition to classic genetic changes in order to adequately assess the true frequency of inactivation of key tumor-suppressor genes, such as *p16*. Occasionally, as for the inactivation of this gene in colon cancer (HERMAN et al. 1995), promoter-region hypermethylation represents the only molecular explanation for loss of tumor-suppressor function in a given tumor type (Fig. 1). In addition to the promoter hypermethylation involving prototype tumor-suppressor genes, another group of probable suppressor genes (Table 2) is also being identified as having loss of function in cancers associated with this promoter-region modification. In some cases, such as the cyclin-kinase-dependent inhibitor gene *p15* in hematopoietic neoplasms (HERMAN et al. 1996a, 1997; DREXLER 1998; MALONEY et al. 1998) and the DNA repair enzyme genes $O^6$-methylguanosyl methyltransferase (*MGMT*) and *glutathione-S-transferase Pi* in multiple tumor types (LEE et al. 1994; ESTELLER et al. 1998b, 1999a), the hypermethylation change is virtually the only molecular lesion that has been associated with loss of gene function.

# 3 Position of CpG-Island Hypermethylation in Tumor Progression

It is clear that, for virtually every tumor type, an accrual of genetic alterations over time is a major force in driving progressive stages of neoplastic development (NOWELL 1976; KINZLER and VOGELSTEIN 1996). What is known about the timing of aberrant hypermethylation events during tumorigenesis? Several recent studies suggest that these DNA alterations can occur quite early in the process. First,

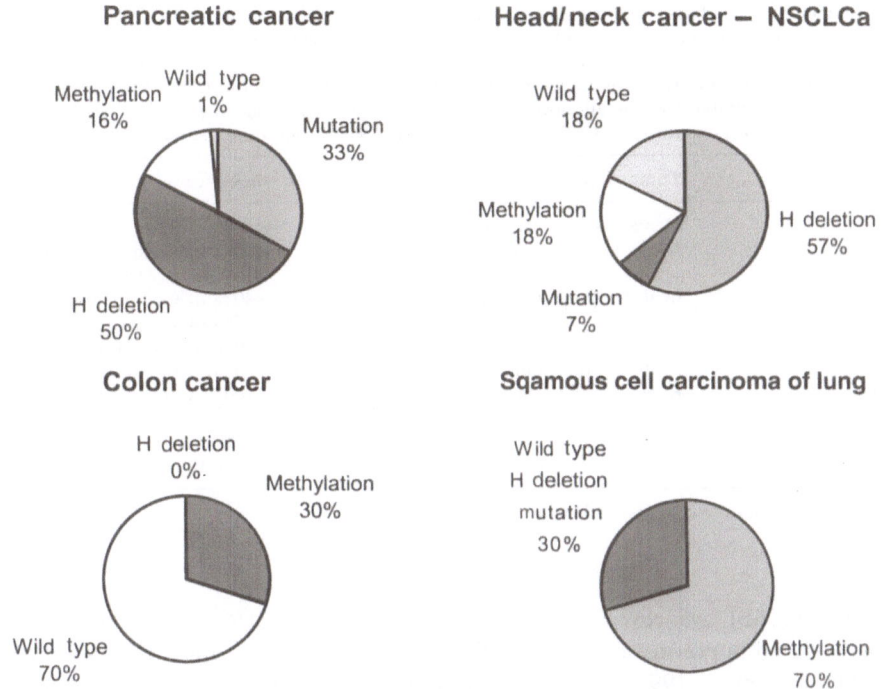

**Fig. 1.** Spectrum of p16$^{INK4a}$ changes in some common forms of human cancer. The different frequencies of p16$^{INK4a}$-inactivating events, as compiled from our own studies and others in the literature, reveal the specificity of how function of this gene is lost in different types of human carcinoma. Note that, in some tumors (such as pancreatic carcinomas), loss in association with promoter hypermethylation is a minor proportion while, in others (such as colon carcinoma), only the epigenetic event appears to be operative. Even within major tumor types, such as non-small-cell lung cancer, the incidence of hypermethylation versus other changes in p16$^{INK4a}$ differs among sub-types. Thus, the incidence of this change in squamous cell lung cancers (*bottom-right chart*) is much higher than when this tumor category is considered as a whole (*top-right chart*). *NSCLCa*, non-small-cell lung carcinoma; *H Deletion*, homozygous deletion

investigations of colon cancer are revealing that some promoter-region hypermethylation patterns in tumors may reflect retention (in the tumor cells) of events that are occurring with increasing frequency in normal colon mucosa as a function of aging (Issa et al. 1994, 1996; Ahuja et al. 1998). These methylation changes appear to involve genes other than those considered to be classic tumor suppressors. Thus, one early form of CpG-island hypermethylation associated with cancer may be a key link between aging and the increased risk of neoplasia, and this is discussed in detail elsewhere in this book.

In addition to the above methylation changes, which appear to be age related, it is also clear that some promoter-region hypermethylation events appear to be specific to cancer cells. These changes involve tumor-suppressor genes, such as those listed in Table 1, and can also occur early in the neoplastic process. Studies employing newly developed and sensitive polymerase chain reaction (PCR) technology for assessment of the methylation status of specific DNA sequences (Her-

**Table 2.** Silencing of candidate tumor-suppressor genes in association with promoter CpG-island methylation. For each of the genes in this table, the same criteria for silencing in association with promoter hypermethylation have been established as those outlined in Table 1. References for *p15*, glutathione-*S*-transferase (*GST*)-*Pi*, and $O^6$-methyl guanosyl methyltransferase (*MGMT*), can be found in the text, and those for tissue inhibitor of metalloproteinases (*TIMP*)-3, death-associated protein (*DAP*)-*kinase*, and *p73* are (KATZENELLENBOGEN et al. 1998; BACHMAN et al. 1999; CORN et al., submitted)

| Gene | Tumor type | Incidence |
|------|-----------|-----------|
| $P15^{INK4a}$ | Hematopoietic neoplasms | ~70% of AML |
| GST-Pi | Multiple solid tumor types | ~90% of Prostate cancer |
| | | ~33% of Breast cancer |
| $O^6$-MGMT | Multiple solid tumor types | ~40% of Colon cancer |
| | | ~15% of Lung cancer |
| | | ~30% of Brain tumors |
| TIMP-3 | Multiple solid tumor types | ~80% of Renal cancer |
| DAP-kinase | B-cell lymphoid malignancies | ~100% Burkitt's lymphoma |
| | | ~80% B-cell lymphoma |
| p73 | T/B cell malignancies | ~60% T-cell ALL |
| | | ~30% Burkitt's lymphoma |

*ALL*, acute lymphoblastic leukemia; *AML*, acute myeloid leukemia

MAN et al. 1996b) are revealing, for example, that the *p16*-gene promoter is hypermethylated in pre-invasive stages of lung and esophageal cancer (WONG et al. 1997; BELINSKY et al. 1998), and the corresponding protein is absent in the lung lesions (BELINSKY et al. 1998). Similarly, *p16*-gene and other promoter-region hypermethylation changes have been detected in the pre-malignant polyps that are thought to represent a precursor stage to colon cancer (SILVERMAN et al. 1989; HERMAN et al. 1995). In addition, *p15*-gene hypermethylation occurs in the myelodyplastic state (QUESNEL et al. 1998; UCHIDA et al. 1998) that can precede leukemia. Experimental systems, such as for mammary epithelium, also indicate that promoter-region hypermethylation may play an early role in progression to cellular immortalization. In these settings, hypermethylation and silencing of the $p16^{INK4A}$ gene accompanies loss of a key mortality checkpoint (FOSTER et al. 1998; HUSCHTSCHA et al. 1998).

Although, it is clear that aberrant promoter-region methylation can occur as an early molecular event in neoplastic progression, much remains to be determined as to the biological effects of this DNA modification at early disease stages. For example, the timing of promoter-region methylation may differ from coding-region mutations as events that provide for full cell-selective advantage during tumor progression. Whereas point mutations occur as a single error in time during a round of DNA replication, multiple cell cycles may be required for a given CpG island to be affected by sufficient methylation density to correlate with maximal silencing of gene transcription (HSIEH 1994; VERTINO et al. 1996). Thus, for a given gene, a range of gene-dosage effects and considerable heterogeneity of cellular and allelic involvement with the methylation changes may be present at any given stage of tumor development. More studies of these dynamics are warranted to understand this process to the fullest.

One particularly exciting concept emerging from studies of the timing and extent of promoter hypermethylation during neoplastic progression is that particular tumor types may evolve patterns of genes involved with this process; these patterns appear to reflect non-random processes of epigenetic instability. In turn, as discussed in detail elsewhere in this book, these epigenetic changes appear to cooperate with genetic alterations to drive the neoplastic process and even to predispose to particular types of mutations. The most vivid illustration of this point comes from studies of colon cancer. A subset of these tumors appears to hypermethylate clusters of genes (including *p16*; AHUJA et al. 1997) simultaneously, and these include the majority of sporadic colon cancers that have the so-called microinstability (MIN+) phenotype (AALTONEN et al. 1993; THIBODEAU et al. 1993). This finding first raised the possibility that the instability of repeat sequences in tumor DNA might be a predisposing factor for development of the hypermethylation changes. However, in a subsequent study, Kolodner and colleagues observed that, in some non-inherited MIN+ colon cancers, the *hMLH1* mismatch-repair gene itself was hypermethylated, and the protein was absent from tumor cells (KANE et al. 1997). A larger collaborative study from our group (HERMAN et al. 1998) and subsequent studies of others (CUNNINGHAM et al. 1998; VEIGL et al. 1998) indeed confirmed these initial findings and suggested that, in as many as 70% of sporadic colorectal cancers with the MIN+ phenotype, the DNA-repeat instability results from loss of mismatch-repair function in association with aberrant methylation of the *hMLH1*-promoter CpG island. In support of this concept, partial reversal of the *hMLH1*-gene methylation changes restores significant MLH1-protein production and substantial mismatch-repair capability in cultured colon and endometrial carcinoma cells (HERMAN et al. 1998). Thus, promoter-region methylation abnormalities can actually proceed and result in important genetic alterations, such as loss of mismatch repair and the resultant MIN+ phenotype.

Other examples of interaction between methylation and genetic changes in cancer are emerging. Several groups have reported that, in cultured brain and other types of tumor cells, loss of MGMT expression is associated with hypermethylation of the gene promoter (COSTELLO et al. 1996; QIAN and BRENT 1997; WATTS et al. 1997). We and others have recently shown that this is also a common change (with correlating loss of protein) in non-cultured tumors, including brain, lung and colon cancers (ESTELLER et al. 1999a; HERFARTH et al. 1999). Most excitingly, this loss of MGMT appears to be a major mechanism for determining the frequency and type of *ras*-gene mutations that occur in colon cancers (manuscript submitted). MGMT activity is required for efficient repair of alkylating damage involving the base guanosines of DNA. This damage, if not repaired, results in G-to-A transitions at the involved sites. Such mutations are the predominant form of *K-ras* mutations in colon cancer, and over 70% of these appear to occur in tumors that harbor hypermethylation of the MGMT gene (manuscript submitted). This situation, then, appears to be another example in which gene-function loss in association with an epigenetic change predisposes to a specific type of genetic alteration.

# 4 What are the Mechanisms Involved in the Gene Silencing Associated with Aberrant Promoter-Region Hypermethylation?

One of the most basic questions regarding the loss of gene function in cancer in association with aberrant promoter-region methylation concerns the precise role of the DNA modification in silencing transcription of the involved genes. Does the methylation initiate the transcription loss? What associated chromatin changes accompany the loss of gene expression? Though the answers to these questions are not yet known, studies of chromatin assembly around methylated promoter templates in experimental systems and around the promoters of genes that are normally silenced in association with hypermethylation are beginning to shed light on this important matter.

As outlined in detail elsewhere in this book, the status of chromatin differs around promoter-region CpG islands that are methylated versus those that are not. The unmethylated islands present in the promoters of most normal genes are surrounded by an open, transcriptionally favorable, nucleosomal arrangement in the presence of highly acetylated histones (ANTEQUERA et al. 1990; TAZI and BIRD 1990; LEWIS and BIRD 1991a,b). In contrast, methylated islands, such as those associated with inactivated genes on the female X-chromosome, are involved with a transcriptionally closed chromatin and de-acetylated histones (ANTEQUERA et al. 1990; JEPPESEN and TURNER 1993). In this setting, the methylation appears to serve as a locking mechanism for transcriptional inactivation, which is initiated by association of a RNA species (termed X-ist) with the X-chromosome genes, which become silenced early in embryogenesis (MARAHRENS et al. 1998).

The chromatin status surrounding the methylated CpG islands in promoters of silenced alleles of imprinted genes on autosomes is probably quite similar to that for genes on the inactive X-chromosome. However, for the imprinted genes, the timing of the methylation events is less well outlined. Methylation of 5′ sites quite distal to the proximal promoter of the imprinted gene, established in germ cells, may serve as the mark that triggers parentally determined silencing of the gene (TILGHMAN 1999). Spread of methylation to CpG islands in the proximal promoters of such genes occurs with actual silencing of the gene during early embryogenesis (TILGHMAN 1999). Whether this methylation initiates loss of gene transcription or locks in such an event following establishment of repressive chromatin is not known.

Another key link between the status of chromatin and the methylation state of DNA is provided by recent discoveries concerning the balance between acetylation of proteins and the presence of proteins that bind preferentially to methylated cytosines. One such protein, methylcytosine-binding protein 2, has been found to participate in a complex that contains active histone de-acetylase activity (JONES et al. 1998; NAN et al. 1998). In experimental settings, specific inhibition of this de-acetylase activity appears to prevent both formation of transcriptionally repressive chromatin around methylated promoter templates and the resultant silencing of transcription (JONES et al. 1998; NAN et al. 1998).

The precise relationship of the above experimental findings for endogenously methylated promoter-region CpG islands, such as those in cancer cells, remains to be determined. For example, it is still unknown whether the hypermethylation initiates the formation of transcriptionally repressive chromatin, as can clearly occur for promoter templates methylated in vitro. However, recent data from our laboratory suggest that promoter-region methylation and histone-deactylase activity do work together to aberrantly silence genes in cancer (CAMERON et al. 1999). We have studied four such genes in cultured tumor cells: *hMLH 1*, *p15*, *p16*, and *TIMP-3*. In each case, for the steady-state situation, the dense CpG-island methylation appeared to be the dominant determinant of the silencing of each gene. Specific inhibition of cellular histone-deactylase activity with the drug trichostatin was ineffective in re-activating any of the hypermethylated genes, while non-methylated genes in the same cells manifested significant upregulation (CAMERON et al. 1999). However, if very modest demethylation was first produced by prior treatment of cells with 5-deoxy-azacytidine (DAC), the trichostatin was subsequently effective in inducing substantial re-expression of each hypermethylated gene (CAMERON et al. 1999). Thus, the processes of promoter methylation and regulation of protein-acetylation status appear to work as layers in neoplastic cells to produce aberrant gene inactivation. The dense methylation may prevent access of protein acetylases to transcription-start sites, thereby overriding the effects of active deacetylase activity. As discussed below, consideration of these dynamics may be most important for any strategies attempting re-activation of hypermethylated genes in cancer cells as a therapeutic maneuver.

# 5  What Mechanisms Underlie the Appearance of Aberrant CpG-Island Methylation in Neoplasia?

One of the critical issues for understanding and potentially reverting the abnormal promoter-region methylation in cancer is understanding the factors leading to this altered DNA modification during tumor progression. Whatever the events prove to be, they must take into account that these regional gains in methylation often occur simultaneously with wide-spread losses of methylation in the genome. This raises the possibility that a fundamental imbalance of the regulation of DNA methylation could be operative in neoplastic cells, thus linking both the gains and losses of methylation found in this setting. Although the precise nature of such a process remains to be determined, clues are emerging regarding the nature of the events that might be involved.

## 5.1  The Potential Role of DNA-Methyltransferase Activity

Although additional mammalian proteins with DNA-methyltransferase (DNA-Mtase) activity have recently been identified (OKANO et al. 1998) and must be evaluated for key roles in establishing methylation patterns in normal and neo-plastic cells, a single such enzyme, DNMT1, appears to be largely responsible for this process (BESTOR et al. 1988; LI et al. 1992; YEN et al. 1992; YODER et al. 1996). A homozygous deletion of the gene encoding this protein in mice produces em-bryonic lethality (LI et al. 1992, 1993), inability of cultured embryonic stem cells to differentiate (BEARD et al. 1995) and profound alterations in gene-imprinting pat-terns (LI et al. 1992, 1993). The enzyme is a 190-kDa protein which resembles bacterial cytosine methylases in its C-terminal third, which contains the catalyti-cally active region. A long N-terminal region, which is well conserved in species as diverse as *Xenopus*, plants and man (TUCKER et al. 1996; YODER et al. 1996), contains regions that help target the enzyme to nuclei and sites of DNA replication (LEONHARDT et al. 1992; CHUANG et al. 1997). This area also modulates the me-thylase activity such that hemi-methylated DNA is the preferred substrate, al-though the enzyme also contains substantial de novo methylating activity (LEONHARDT and BESTOR 1993; PRADHAN et al. 1997).

A substantial body of data has suggested that increases in DNMT1 activity may play a role in cellular transformation, although proof of this remains to be established. Increases in expression of the gene have been observed at the mes-senger-RNA and protein levels in neoplastic cells of multiple types (EL-DEIRY et al. 1991; ISSA et al. 1993; LEE et al. 1996; SCHMUTTE et al. 1996). When the role of cell replication is taken into account, these increases are modest (in the range of two- to threefold relative to increases in normal cells; ISSA et al. 1993; LEE et al. 1996; SCHMUTTE et al. 1996). However, NIH-3T3 cells engineered to overexpress the murine DNMT1 gene to a similar degree can produce experimental cell transfor-mation (WU et al. 1993). More recently, such increases have been suggested to play such a role in cells responding to increased *ras*-gene activity (MACLEOD et al. 1995) and over-expressing the *c-fos* oncogene (BAKIN and CURRAN 1999). In these settings of cell engineering, the increased DNMT1 activity results in increases in overall genomic methylation (WU et al. 1993; MACLEOD et al. 1995; BAKIN and CURRAN 1999) and, in human fibroblasts, a time-dependent hypermethylation of some promoter-region CpG islands (VERTINO et al. 1996).

Thus, there is a growing body of data implicating increases in DNMT1 activity in the process of cellular transformation. However, it is virtually certain that such increases act as events complimentary to other cell alterations critical for full cel-lular transformation. All of the cell types involved in the studies cited above were immortalized prior to experimental increases in DNMT-1 activity. Furthermore, in studies of cell fusion between transformed and normal cells, the cell hybrids maintain the high DNA-Mtase activity of the parent transformed cells even though these fusion cells are initially immortalized but not transformed (KUERBITZ and BAYLIN 1996). Further studies, particularly those employing genetic manipulation to alter enzyme activity, are clearly warranted to fully decipher the role of DNA-Mtase in the evolution of neoplasia. Also, understanding whether such a role can be explained primarily through the hypermethylation and gene-silencing events, which are the focus of this review, remains to be determined. Finally, it is our belief that the increased DNA-Mtase activity during tumor progression may be a very important permissive event that cooperates with local changes that render CpG islands abnormally vulnerable to methylation. As discussed below, such local changes could involve alterations in the structures of proteins or in their post-translational modification, which are critical to the transcription process. Particularly important may be the chromatin-modeling proteins (that modulate but do not solely control) the transcriptional process. Such proteins may also be critical for targeting DNA-Mtases to DNA such that certain areas are protected from methylation and others are normally methylated. To date, only the protein, proliferating cell nuclear antigen (CHUANG et al. 1997), has been linked to localization of DNMT1 to DNA. This may be relevant in targeting the enzyme to DNA replication foci rather than in determining which regions of the genome are methylated. Loss or abnormal gain of key chromatin-modeling functions in tumor cells could thus underlie the simultaneous gains and losses of DNA methylation that are now apparent as hallmarks of the neoplastic state.

An additional factor that must be considered in the protection of CpG islands from methylation is the role of proteins with demethylase activity. Such enzymes have long been sought, and recent cloning of a methylcytosine-binding protein that appears to contain demethylase activity has been an exciting advance (BHATTA-CHARYA et al. 1999; CEDAR and VERDINE 1999). This protein, which has also been identified as MBD2b (HENRICH and BIRD 1998), may participate in modulating the genome-wide patterns of DNA methylation. Its role in normal and neoplastic cells must be investigated further.

## 5.2 Do Alterations in Local Control of Chromatin Assembly Mediate Abnormal Promoter-Region Methylation in Neoplasia?

At present, there is little direct evidence to outline what local mechanisms might allow the preponderance of promoter CpG islands on autosomal chromosomes to be protected from methylation in normal cells and what processes might be altered in neoplasia to render these regions vulnerable to this DNA modification. However, hints are emerging, at least for some gene loci, that implicate changes in proteins that directly or indirectly regulate the process of gene transcription. For example, in transient-transfection assays, where methylation of the introduced construct should not be a factor, breast and prostate cancer cells that harbor both an endogenously hypermethylated E-cadherin (*E-cad*) promoter and abnormal silencing of the gene do not activate an exogenous *E-cad* promoter with the same efficiency as do the same types of cells, in which the endogenous gene promoter is unmethylated and highly expressed (GRAFF et al. 1995). These findings suggest that *trans*-acting factors, which modulate (but do not fully promote) transcription of the *E-cad* gene could be lost in some tumors. In this setting, a partial loss of transcription capacity for the gene may be a predisposing factor for the promoter CpG island to become methylated and subsequently to become fully inactivated. It is not yet clear whether such dynamics are operative at other aberrantly hypermethylated genes in tumor cells.

If local factors are critical for protection of promoter CpG islands from methylation, what types of proteins might be involved? In this regard, the structure of the CpG islands and the relationship of these to the patterns of methylation around them in normal cells need to be considered. Many genes are similar in these parameters in that their islands, averaging 1–2kb in length, span the transcription-start site and extend through the first intron. At the borders of such islands, particularly on the 5′ side, highly methylated sequences, such as Alu repeats, are often found (GRAFF et al. 1997). Such repeats have been proposed as actual initiating sites for de novo methylation, which can spread bi-directionally from the first CpG dinucleotides that are modified (MUMMANENI et al. 1995). The CpG islands residing nearby must then be protected from the spread of the methylation. One candidate for such protection is the CpG-containing transcription-binding site for the transcription factor Sp1. These sequences are often clustered near the 5′ and 3′ borders of the islands (GRAFF et al. 1997) and have been shown experimentally, in at least one gene, to be involved in protection of the island from methylation (BRANDEIS et al. 1994; MACLEOD et al. 1994; MUMMANENI et al. 1995). However, mice with homozygous disruption of the Sp1 gene have normal, unmethylated CpG islands at all gene loci tested (MARIN et al. 1997). Thus, it seems clear that, although the Sp1-binding sites may normally be important for protecting CpG islands from methylation, proteins other than Sp1 itself (perhaps other members of this family of transcription factors or partners with which they interact) may be involved.

One of the most exciting possibilities for envisioning proteins that might protect CpG islands from methylation might be derived from considering these regions as three-dimensional domains that could involve looped structures (maintained by chromatin proteins) that are inaccessible to DNMT1. The multi-protein complexes regulating gene transcription are envisioned, of course, to be arranged in such domains (CAREY 1998; PAREKH and MANIATIS 1999). Stable maintenance of such active or repressive transcription domains, which involve higher-order chromatin structures in which the basal transcription machinery may be embedded, would be excellent candidates for modulating the methylation status of promoter regions. Such higher-order structures involve chromatin-modeling protein complexes, such as those formed by the switching/sucrose-non-fermenting protein (SWI/SNF) family (PETERSON 1996; WANG 1996; POLLARD and PETERSON 1998), and transcriptional domains maintained by multi-protein complexes, such as those regulating the interferon (IFN)-β promoter (PAREKH and MANIATIS 1999). Loss of a single protein from such complexes would not be expected to abolish the transcriptional capacity of a gene but, as for the IFN-β promoter (PAREKH and MANIATIS 1999), may lead to partial defects in function, such as those discussed earlier for the *E-cad* gene. While the effects of the protein loss on transcription may be partial, the consequences for maintenance of chromatin structure could be profound and, perhaps, could render the involved region susceptible to loss of protection from methylation.

An exciting example of the involvement of proteins in establishing genomic methylation patterns has recently been provided by a mutation model in the plant

*Arabidopsis*. This mutation leads simultaneously to wide-spread loss of genomic methylation and gain of methylation in isolated gene promoters, with concomitant loss of gene function (PASZKOWSKI and SCHEID 1998). This situation is, of course, not unlike the pattern discussed above for cancer cells. Recently, Richards and colleagues defined the gene mutation leading to the methylation alterations in *Arabidopsis* as encoding a protein with high homology to the above-mentioned SWI/SNF family of chromatin-modeling proteins (JEDDELOH et al. 1999). This was the first demonstration that such proteins may be integral to establishing genomic methylation patterns. Another chromatin-modeling protein, SNF5, has recently been found to be mutated in an uncommon form of pediatric tumor (VERSTEEGE et al. 1998), but the ramifications for methylation patterns, if any, are not known.

In summary, understanding the mechanisms responsible for normal protection of most CpG islands from methylation and how these mechanisms become lost during neoplastic evolution is a critical priority for tumor-biology research. A role for increased activity of DNMT1 is highly probable, and the participation of other recently cloned DNA-Mtases must be sought. The exciting possibility that proteins (especially those with chromatin-remodeling activity) participate in transcriptional control must be vigorously pursued. Elucidation of the above points should prove an exciting chapter in cancer biology and may prove important for new approaches to the diagnosis, prevention, and management of tumors, as discussed further below.

# 6 The Clinical Significance of DNA Hypermethylation in Cancer

## 6.1 Implications for Therapy

As discussed in the sections above, a growing number of genes important for tumor development and progression are being found to be inactivated in neoplastic cells, in association with aberrant promoter-region methylation. Since the transcriptional repression associated with this process is potentially reversible with demethylating agents, such as DAC, the re-activation of the involved genes is a therapeutic target that has received considerable recent attention. In fact, DAC was developed as a potential chemotherapy agent, with some responses documented predominantly in the hematopoietic malignancies (GATTEI et al. 1993; MANDELLI 1993; PETTI et al. 1993; SILVERMAN et al. 1993; WILLEMZE et al. 1993; ZAGONEL et al. 1993). The drug continues to undergo clinical trials in leukemia and the myelodysplastic disorders. However, DAC has an inherent toxicity that has been a problem, and this activity may well be unrelated to the hypomethylating properties of the drug. DAC is effective in blocking DNA-Mtase activity only when it is incorporated into DNA as a base. In this setting, the drug covalently binds the enzyme into DNA, where it has the potential to act as an adduct with DNA-damaging properties (JUTTERMANN et al. 1994; FERGUSON et al. 1997). In fact, this may account for the finding that the cytotoxicity observed for DAC appears to be directly proportional to the level of DNA-Mtase activity in a given cell type (JUTTERMANN et al. 1994; FERGUSON et al. 1997).

Despite the above problems, much may still be learned from ongoing trials with DAC in hematopoietic tumors. A fundamental question is whether tumor response is temporally related to the re-activation of key genes that are hyper-methylated in the pre-treatment state. The myelodyplastic disorders and leukemias represent ideal situations in which to ask this question, since timed samples can be obtained in a relatively non-invasive manner after therapy. Such studies are on-going and should provide important answers regarding both the extent to which transcriptionally repressed genes can be reactivated in vivo and the consequences of such reactivation for anti-tumor effects.

The ultimate value of exploiting promoter hypermethylation as a therapeutic target may depend on the development of strategies other than the use of DAC to block or reduce the methylation process. Several such approaches, including use of antisense molecules targeted against DNMT1, are being tested in cell cultures and animal models (MacLEOD and SZYF 1995). These studies have suggested that anti-transformation and tumor effects can be achieved, and the results continue to encourage targeting of DNA-Mtase activity as a therapeutic maneuver. Other approaches that could have even greater potential include the design of small molecules that may block the enzyme activity without incorporation into DNA. Finally, as discussed previously, DNA methylation and regulation of protein-acetylation status may work together at transcription-start sites to influence gene-expression status in cancer cells. Strategies to simultaneously inhibit the met-hylation and histone-deacetylase processes may be the most exciting for achieving re-activation of key genes for therapeutic purposes.

## 6.2 Hypermethylated Loci as Biomarkers for Tumors

One primary goal of applying our growing understanding of the molecular changes in cancer, particularly as it relates to DNA changes, is the development of new early-detection and prognosis monitoring strategies. It has become apparent over the past one or two years that detection of hypermethylated loci may offer one of the most promising current approaches to this goal. These modified DNA se-quences meet multiple criteria for an optimal tumor biomarker. First, as discussed in the sections above, hypermethylated CpG-island sequences are associated with virtually every type of human cancer, and many of these changes appear to be tumor specific. Second, as discussed in earlier sections, every human cancer type that has been studied harbors multiple different hypermethylated loci, and early studies suggest that a number of such markers, used as a tumor-specific panel, could mark each cancer. Third, PCR-based assays for assessment of the met-hylation status of specific sequences have been developed and may have the re-quired sensitivity to detect (in tissues such as serum) aberrantly hypermethylated loci associated with tumor DNA. The methylation-specific PCR assay, as routinely used in our laboratory programs, can detect one methylated allele among 1000 unmethylated alleles, even in DNA from paraffin-embedded specimens (HERMAN et al. 1996b). Fourth, many specific genetic changes in cancer, such as point mutations in gene coding regions, occur within a broad genomic region encompassing a given gene. Thus, one must know the specific mutation in the tumor DNA of a given patient to efficiently develop an assay that will detect the change in DNA taken from distant sites. In contrast, CpG-island hypermethylation always occurs at a defined region within the promoter of a gene. Thus, these methylation changes can be detected in samples from peripheral sites – in all patients – by a single PCR strategy, without sampling the primary tumor. Finally, in the PCR assays, positive results are detected by appearance, rather than the loss, of a signal. Thus, background normal DNA poses much less of a problem than it does for detection of markers, such as loss of allelic heterozygosity.

The proof of principle for use of hypermethylated loci as tumor biomarkers is already emerging in small studies aimed at testing the hypotheses put forth above. Thus, an aberrantly hypermethyated $p16$ promoter has been detected in the sputum of patients at high risk for developing lung cancer (BELINSKY et al. 1998). Most impressively, in serum from patients with lung cancer (ESTELLER et al. 1999b) and those with hepatocellular carcinoma (WONG et al. 1999), hypermethylation of the $p16$ promoter and the promoters of other genes can be detected with 100% specificity and 70–80% sensitivity. Some of the patients with lung cancer had stage-I disease, which represents a potential surgically curative status. The above types of studies are only a start and must be vastly extended. Critical issues, including identifying the determinants of when serum (or other types of peripheral samples) will be positive for the methylation changes and the role of tumor burden, must be investigated. However, the above initial findings are extremely encouraging and justify a vigorous effort to establish whether detection of hypermethylated CpG islands can provide sensitive biomarkers useful for monitoring tumor-prevention strategies and establishing assays for early cancer detection and for refined approaches to gauging prognosis.

# 7 Summary

In summary, it is apparent that alterations in DNA methylation are a fundamental molecular change associated with the neoplastic process and have important biologic implications for tumor initiation and progression. The promoter-region hypermethylation events covered in the present chapter are especially critical and can frequently serve as alternative mechanisms for coding-region mutations for loss of key gene function in neoplastic cells. The mechanisms underlying the precise role of this hypermethylation in gene silencing must be further defined, as must the determinants of the hypermethylation changes themselves. The therapeutic implications of promoter-region hypermethylation must be explored, and a potential use for establishing this change as a sensitive biomarker for use in multiple types of cancer-risk assessment and detection assays has already emerged. The next few years should see exciting advances in our understanding of an epigenetic process which, in conjunction with genetic alterations, appears to drive the process of neoplasia.

# References

Aaltonen LA, Peltomaki P, Leach FS, Sistonen P, Pylkkanen L, Mecklin JP, Jarvinen H, Powell SM, Jen J, Hamilton SR (1993) Clues to the pathogenesis of familial colorectal cancer. Science 260:812–816

Ahuja A, Mohan AL, Li Q, Stolker JM, Herman JG, Hamilton SR, Baylin SB, Issa J-P (1997) Association between CpG island methylation and microsatellite instability in colorectal cancer. Cancer Res 57:3370–3374

Ahuja A, Li Q, Mohan AL, Baylin SB, Issa J-P (1998) Aging and DNA methylation in colorectal mucosa and cancer. Cancer Research 58:5489–5494

Antequera F, Bird A (1993) CpG islands. In: DNA methylation: molecular biology and biological significance, Jost JP, Saluz HP (eds) (Basel: Birkhauser Verlag), pp 169–185

Antequera F, Boyes J, Bird A (1990) High levels of de novo methylation and altered chromatin structure at CpG islands in cell lines. Cell 62:503–514

Bachman KE, Herman JG, Corn PG, Merlo A, Costello JF, Cavenee WK, Baylin SB (1999) Methylation-associated silencing of the tissue inhibitor of metalloproteinase-3 gene suggests a suppressor role in kidney, brain and other human cancers. Cancer Res 59:798–802

Bakin AV, Curran T (1999) Role of DNA 5-methylcytosine transferase in cell transformation by fos. Science 283:387–390

Baylin SB, Herman JG, Graff JR, Vertino PM, Issa J-P (1998) Alterations in DNA methylation: a fundamental aspect of neoplasia: In Advances in Cancer Research, Klein G, Van de Woude GF (eds) (San Diego: Academic Press), pp 141–196

Beard C, Li E, Jaenisch R (1995) Loss of methylation activates Xist in somatic but not in embryonic cells. Genes and Dev 9:2325–2334

Belinsky SA, Nikula KJ, Palmisano WA, Michels R, Saccomanno G, Gabrielson E, Baylin S, Herman JG (1998) Aberrant methylation of p16$^{INK4a}$ is an early event in lung cancer and a potential biomarker for early diagnosis. Proc. Natl Acad Sci USA 95:11891–11896

Bestor T, Laudano A, Mattaliano R, Ingram V (1988) Cloning and sequencing of a cDNA encoding DNA methyltransferase of mouse cells. J Mol Biol 203:971–983

Bhattacharya SK, Ramchandani S, Cervoni N, Szyf M (1999) A mammalian protein with specific demethylase activity for mCpG DNA. Nature 397:579–583

Bird AP (1987) CpG islands as gene markers in the vertebrate nucleus. Trends Genet. 3

Bird AP (1992) The essentials of DNA methylation. Cell 70:5–8

Brandeis M, Frank D, Keshet I, Siegfried Z, Mendelsohn M, Nemes A, Temper V, Razin A, Cedar H (1994) Sp1 elements protect a CpG island from de novo methylation. Nature 29:435–438

Cameron EE, Bachman KE, Myohanen S, Herman JG, Baylin SB (1999) Synergy of demethylation and histone deacetylase inhibition in the re-expression of genes silenced in cancer. Nature Genetics 21:103–107

Carey M (1998) The enhanceosome and transcriptional synergy. Cell 92:5–8

Cedar H (1988) DNA methylation and gene activity. Cell 53:3–4

Cedar H, Verdine GL (1999) The amazing demethylase. Nature 397:568–569

Chuang LS, Ian HI, Koh TW, Ng HH, Xu G, Li BF (1997) Human DNA–(cytosine-5) methyltransferase–PCNA complex as a target for p21WAF1. Science 277:1996–2000

Costello JF, Futscher BW, Tano K, Graunke DM, Pieper RO (1996) Graded methylation in the promoter and in the body of the $O^6$-methylguanine-DNA methyltransferase gene correlates with MGMT expression in human glioma cells. Cancer Res 56:13916 13924

Cunningham JM, Christensen ER, Tester DJ, Kim C-Y, Roche PC, Burgart LJ, Thibodeau SN (1998) Hypermethylation of the hMLH1 promoter in colon cancer with microsatellite instability. Cancer Res 58:3455–3460

Drexler HG (1998) Review of alterations of the cyclin-dependent kinase inhibitor INK4 family genes *p15*, *p16*, *p18*, and *p19* in human leukemia-lymphoma cells. Leukemia 12:845–859

El-Deiry WS, Nelkin BD, Celano P, Yen R-WC, Falco JP, Hamilton SR, Baylin SB (1991) High expression of the DNA methyltransferase gene characterizes human neoplastic cells and progression stages of colon cancer. Proc Natl Acad Sci USA 88:3470–3474

Esteller M, Levine R, Baylin SB, Ellenson LH, Herman JG (1998a) MLH1-promoter hypermethylation is associated with the microsatellite instability phenotype in sporadic endometrial carcinomas. Oncogene 16:2413–2417

Esteller M, Corn PG, Urena JM, Gabrielson E, Baylin SB, Herman JG (1998b) Inactivation of glutathione *S*-transferase P1 gene by promoter hypermethylation in human neoplasia. Cancer Res 58:4515–4518

Esteller M, Hamilton SR, Burger PC, Baylin SB, Herman JG (1999a) Inactivation of the DNA-repair gene $O^6$-methylguanine–DNA methyltransferase by promoter hypermethylation is a common event in primary human neoplasia. Cancer Res 59:793–797

Esteller M, Sanchez-Cespedes M, Rossel R, Sidransky D, Baylin SB, Herman JG (1999b) Detection of aberrant promoter hypermethylation of tumor-suppressor genes in serum DNA from non-small-cell lung cancer patients. Cancer Res 59:67–70

Ferguson AT, Vertino PM, Spitzner JR, Baylin SB, Muller MT, Davidson NE (1997) Role of estrogen receptor gene demethylation and DNA-methyltransferase DNA adduct formation in 5-aza-2'-deoxycytidine-induced cytotoxicity in human breast cancer cells. J Biol Chem 272:32260–32266

Fleisher AS, Esteller M, Wang S, Tamura G, Suzuki H, Yin J, Zou T-T, Abraham JM, Kong D, Smolinski KN, Shi Y-Q, Rhyu M-G, Powell SM, James SP, Wilson KT, Herman JG, Meltzer SJ (1999) Hypermethylation of the hMLH1 gene promoter in human gastric cancers with microsatellite instability. Cancer Research 59:1090–1095

Foster SA, Wong DJ, Barrett MT, Galloway DA (1998) Inactivation of p16 in human mammary epithelial cells by CpG island methylation. Mol Cell Biol 18:1793–1801

Gattei V, Aldinacci D, Petti MC, Da Ponte A, Zagonel V, Pinto A (1993) In vitro and in vivo effects of 5-aza-2'-deoxycytidine (decitabine) on clonogenic cells from acute myeloid leukemia patients. Leukemia 7:42–48

Gonzalez-Zulueta M, Bender CM, Yang AS, Nguyen T, Beart RW, Van Tornout JM, Jones PA (1995) Methylation of the 5' CpG island of the *p16/CDKN2* tumor-suppressor gene in normal and transformed human tissues correlates with gene silencing. Cancer Res 55:4531–4535

Graff JR, Herman JG, Lapidus RG, Chopra H, Xu R, Jarrard DF, Isaacs WB, Pitha PM, Davidson NE, Baylin SB (1995) E-cadherin expression is silenced by DNA hypermethylation in human breast and prostate carcinomas. Cancer Res 55:5195–5199

Graff JR, Herman JG, Myohanen S, Baylin SB, Vertino PM (1997) Mapping patterns of CpG-island methylation in normal and neoplastic cells implicates both upstream and downstream regions in de novo methylation. J Biol Chem 272:22322–22329

Graff JR, Greenberg VE, Herman JG (1998) Distinct patterns of E-cadherin CpG-island methylation in papillary, follicular, Hurthle's cell, and poorly differentiated human thyroid carcinoma. Cancer Res 58:2063–2066

Henrich B, Bird A (1998) Identification and characterization of a family of mammalian methyl-CpG binding proteins. Mol Cell Biol 18:6538–6547

Herfarth KK, Brent TP, Danam RP, Remack JS, Kodner IJ, Wells SAJ, Goodfellow PJ (1999) A specific CpG-methylation pattern of the MGMT promoter region associated with reduced MGMT expression in primary colorectal cancers. Mol Carcinog 24:90–98

Herman JG, Latif F, Weng Y, Lerman MI, Zbar B, Liu S, Samid D, Duan D-SR, Gnarra JR, Linehan WM, Baylin SB (1994) Silencing of the VHL tumor-suppressor gene by DNA methylation in renal carcinoma. Proc Natl Acad Sci USA 91:9700–9704

Herman JG, Merlo A, Mao L, Lapidus RG, Issa J-P, Davidson NE, Sidransky D, Baylin SB (1995) Inactivation of the *CDKN2/p16/MTS1* gene is frequently associated with aberrant DNA methylation in all common human cancers. Cancer Res 55:4525–4530

Herman JG, Jen J, Merlo A, Baylin SB (1996a) Hypermethylation-associated inactivation indicates a tumor-suppressor role for *p15(INK4B)* Cancer Res 56:722–727

Herman JG, Graff JR, Myohanen S, Nelkin BD, Baylin SB (1996b) Methylation-specific PCR: a novel PCR assay for methylation status of CpG islands. Proc Natl Acad Sci USA 93:9821 9826

Herman JG, Civin CI, Issa J-PJ, Collector MI, Sharkis SJ, Baylin SB (1997) Distinct patterns of inactivation of p15$^{INK4B}$ and p16$^{INK4A}$ characterize the major types of hematological malignancies. Cancer Res 57:837–841

Herman JG, Umar A, Polyak K, Graff JR, Ahuja N, Issa J-PJ, Markowitz S, Willson JK, Hamilton SR, Kinzler KW, Kane MF, Kolodner RD, Vogl S, Kunkel TA, Baylin SB (1998) Incidence and functional consequences of hMLH1-promoter hypermethylation in colorectal carcinoma. Proc Nat Acad Sci USA 95:6870–6875

Hsieh C-L (1994) Dependence of transcriptional repression on CpG methylation density. Mol and Cell Biol 14:5487–5494

Huschtscha LI, Noble JR, Neumann AA, Moy EL, Barry P, Melki JR, Clark SJ, Reddel RR (1998) Loss of p16INK4 expression by methylation is associated with life-span extension of human mammary epithelial cells. Cancer Res 58:3508–3512

Issa JP, Vertino PM, Wu J, Sazawal S, Celano P, Nelkin BD, Hamilton SR, Baylin SB (1993) Increased cytosine–DNA-methyltransferase activity during colon cancer progression. J Natl Cancer Inst 85:1235–1240

Issa JP, Ottaviano YL, Celano P, Hamilton SR, Davidson NE, Baylin SB (1994) Methylation of the oestrogen receptor CpG island links ageing and neoplasia in human colon. Nature Genetics 7:536–540

Issa JP, Vertino PM, Boehm CD, Newsham IF, Baylin SB (1996) Switch from monoallelic to biallelic human IGF2-promoter methylation during aging and carcinogenesis. Proc Natl Acad Sci USA 93:11757–11762

Jeddeloh JA, Stokes TL, Richards EJ Maintenance of genomic methylation requires a SWI2/SNF2-like protein. Nature Genet 22:94–97

Jeppesen P, Turner BM (1993) The inactive X chromosome in female mammals is distinguished by a lack of histone H4 acetylation, a cytogenetic marker for gene expression. Cell 74:281–289

Jones PA (1996) DNA-methylation errors and cancer. Cancer Res. 56:2463–2467

Jones PA, Laird PW (1999) Cancer epigenetics comes of age. Nature Genetics 21:163–167

Jones PL, Veenstra GJC, Wade PA, Vermaak D, Kass SU, Landsberger N, Strouboulis J, Wolffe AP (1998) Methylated DNA and MeCP2 recruit histone deacetylase to repress transcription. Nature Genet 19, 187

Juttermann R, Li E, Jaenisch R (1994) Toxicity of 5-aza-2'-deoxycytidine to mammalian cells is mediated primarily by covalent trapping of DNA methyltransferase rather than DNA demethylation. Proc Natl Acad Sci USA 91:11797–11801

Kafri T, Gao X, Razin A (1993) Mechanistic aspects of genome-wide demethylation in the preimplantation mouse embryo. Proc Natl Acad Sci USA 90:10558–10562

Kamb A, Gruis NA, Weaver-Feldhaus J, Liu Q, Harshman K, Tavtigian SV, Stockert E, Day RS, Johnson BE, Skolnick MH (1994) A cell-cycle regulator potentially involved in genesis of many tumor types. Science 264:436–439

Kane MF, Loda M, Gaida GM, Lipman J, Mishra R, Goldman H, Jessup JM, Kolodner R (1997) Methylation of the hMLH1 promoter correlates with lack of expression of hMLH1 in sporadic colon tumors and mismatch repair-defective human tumor cell lines. Cancer Res 57:808–811

Katzenellenbogen RA, Baylin SB, Herman JG. Hypermethylation of the DAP-kinase CpG island is a common alteration in B-cell malignancies. Blood, In Press

Kinzler KW, Vogelstein B (1996) Lessons from hereditary colorectal cancer. Cell 87:159–170

Kuerbitz SJ, Baylin SB (1996) Retention of unmethylated CpG island alleles in human diploid fibroblast × fibrosarcoma hybrids expressing high levels of DNA methyltransferase. Cell Growth and Diff 7:847–853

Lee PJ, Washer LL, Law DJ, Boland CR, Horon IL, Feinberg AP (1996) Limited up-regulation of DNA methyltransferase in human colon cancer reflecting increased cell proliferation. Proc Natl Acad Sci USA 93:10366–10370

Lee WH, Morton RA, Epstein JI, Brooks JD, Campbell PA, Bova GS, Hsieh WS, Isaacs Wb, Nelson WG (1994) Cytidine methylation of regulatory sequences near the π-class glutathione S-transferase gene accompanies human prostatic carcinogenesis. Proc Nat Acad Sci USA 91:11733–11737

Leonhardt H, Bestor TH (1993) Structure, function and regulation of mammalian DNA methyltransferase. In DNA methylation: molecular biology and biological significance, Jost JP, Saluz HP (eds) (Basel: Birkhauser Verlag), pp 109–119

Leonhardt H, Page AW, Weier H-U, Bestor TH (1992) A targeting sequence directs DNA methyltransferase to sites of DNA replication in mammalian nuclei. Cell 71:865–873

Leung SY, Yuen ST, Chung LP, Chu KM, Chan AS, Ho JC (1999) hMLH1 promoter methylation and lack of hMLH1 expression in sporadic gastric carcinomas with high-frequency microsatellite instability. Cancer Res 59:159–164

Lewis J, Bird A (1991) DNA methylation and chromatin structure. FEBS Lett. 285:155–159

Li E, Bestor TH, Jaenisch R (1992) Targeted mutation of the DNA methyltransferase gene results in embryonic lethality. Cell 69:915–926

Li E, Beard C, Jaenisch R (1993) Role for DNA methylation in genomic imprinting. Nature 366:362–365

MacLeod AR, Szyf M (1995) Expression of antisense to DNA methyltransferase mRNA induces DNA demethylation and inhibits tumorigenesis. J Biol Chem 270:8037–8043

MacLeod AR, Rouleau J, Szyf M (1995) Regulation of DNA methylation by the Ras signaling pathway. J Biol Chem 270:11327–11337

MacLeod D, Charlton J, Mullins J, Bird AP (1994) Sp1 sites in the mouse *aprt* gene promoter are required to prevent methylation of the CpG island. Genes and Dev. 8:2282–2292

Maloney KW, McGavran L, Odom LF, Huger SP (1998) Different patterns of homozygous p16$^{INK4a}$ and p15$^{INK4B}$ deletions in childhood acute lymphoblastic leukemias containing distinct E2A translocations. Leukemia 12:1417–1421

Mandelli F (1993) Introduction to the workshop on DNA methyltransferase inhibitors. Leukemia 7:1–2

Marahrens Y, Loring J, Jaenisch R (1998) Role of the Xist gene in X chromosome choosing. Cell 92: 657–664

Mareel M, Bracke M, Van Roy F (1995) Cancer metastasis: negative regulation by an invasion-suppressor complex. Cancer Detection and Prev 19:451–464

Marin M, Karis A, Visser P, Grosveld F, Philipsen S (1997) Transcription factor Sp1 is essential for early embryonic development but dispensable for cell growth and differentiation. Cell 619–628

Merlo A, Herman JG, Mao L, Lee DJ, Gabrielson E, Burger PC, Baylin SB, Sidransky D (1995) 5' CpG island methylation is associated with transcriptional silencing of the tumour suppressor p16/CDKN2/MTS1 in human cancers. Nature Med 1:686–692

Modrich P, Lahue R (1996) Mismatch repair in replication fidelity genetic recombination, and cancer biology. Annu Rev Biochemistry 65:101–133

Mummaneni P, Walker KA, Bishop PL, Turker MS (1995) Epigenetic gene inactivation induced by a *cis*-acting methylation center. J Biol Chem 270:788–792

Myohanen SK, Baylin SB, Herman JG (1998) Hypermethylation can selectively silence individual p16ink4A alleles in neoplasia. Cancer Res 58:591–593

Nan X, Ng H-H, Johnson CA, Laherty CD, Turner BM, Eisenman RN, Bird A (1998) Transcriptional repression by the methyl-CpG-binding protein MeCP2 involves a histone deacetylase complex. Nature 393, 386

Nowell PC (1976) The clonal evolution of tumor cell populations. Science 194:23–28

Okano M, Xie S, Li E (1998) Cloning and characterization of a family of novel mammalian DNA (cytosine-5) methyltransferases. Nature Genet 19:219–220

Parekh BS, Maniatis T (1999) Virus infection leads to localized hyperacetylation of histones H3 and H4 at the IFN-β promoter. Molecular Cell 3:125–129

Paszkowski J, Scheid OM (1998) Plant genes: the genetics of epigenetics. Current Biology 8, R206–R208

Peterson CL (1996) Multiple SWItches to turn on chromatin? Curr Opin Genet Dev 6:171–175

Petti MC, Mandelli F, Zagonel V, De Gregoris C, Merola MC, Latagliata R, Gattei V, Fazi P, Monfardini S, Pinto A (1993) Pilot study of 5-aza-2'-deoxycytidine (decitabine) in the treatment of poor prognosis acute myelogenous leukemia patients: preliminary results. Leukemia 7:36–41

Pollard KJ, Peterson CL (1998) Chromatin remodeling: a marriage between two families? Bioessays 20:771–780

Pradhan S, Talbot D, Sha M, Benner J, Hornstra L, Li E, Jaenisch R, Roberts RJ (1997) Baculovirus-mediated expression and characterization of the full-length murine DNA methyltransferase. Nucleic Acids Research 25:4666–4673

Qian XC, Brent TP (1997) Methylation hot spots in the 5' flanking region denote silencing of the $O^6$-methylguanine–DNA-methyltransferase gene. Cancer Res 57:3672–3677

Quesnel B, Guillerm G, Vereecque R, Wattel E, Preudhomme C, Bauters F, Vanrumbeke M, Fenaux P (1998) Methylation of the p15$^{INK4b}$ gene in myelodysplastic syndromes is frequent and acquired during disease progression. Blood 91:2985–2990

Schmutte C, Yang AS, Nguyen TT, Beart RW, Jones PA (1996) Mechanisms for the involvement of DNA methylation in colon carcinogenesis. Cancer Res 56:2375–2381

Serrano M, Hannon GJ, Beach D (1993) Nature 366:704

Serrano M, Lee HW, Chin L, Cordoncardo C, Beach D, DePinho RA (1996) Role of the INK4A locus in tumor suppression and cell mortality. Cell 85:27–37

Sherr CJ (1994) G1-phase progression: cycling on cue. Cell 79:551–555

Silverman AL, Park JG, Hamilton SR, Gazdar AF, Luk GD, Baylin SB (1989) Abnormal methylation of the calcitonin gene in human colonic neoplasms. Cancer Res 49:3468–3473

Silverman LR, Holland JF, Weinberg RS, Alter BP, Davis RB, Ellison RR, Demakos EP, Cornell CJ, Carey RW, Schiffer C, Frei E, McIntyre OR (1993) Effects of treatment with 5-azacytidine on the in vivo and in vitro hematopoiesis in patients with myelodysplastic syndromes. Leukemia 7:21–29

Simpkins SB, Bocker T, Swisher EM, Mutch DG, Gersell DJ, Kovatich AJ, Palazzo JP, Fishel R, Goodfellow PJ (1999) MLH1 promoter methylation and gene silencing is the primary cause of microsatellite instability in sporadic endometrial cancers. Hum Mol Genet 8:661–666

Strauss M, Lukas J, Bartek J (1995) Unrestricted cell cycling and cancer. Nature Med 1:1245–1246

Tazi J, Bird A (1990) Alternative chromatin structure at CpG islands. Cell 60:909–920

Thibodeau SN, Bren G, Schaid D (1993) Microsatellite instability in cancer of the proximal colon. Science 260:816–819

Thomas DC, Umar A, Kunkel TA (1996) Microsatellite instability and mismatch repair defects in cancer. Mutat Res 350:201–205

Tilghman SM (1999) The sins of fathers and mothers: Genomic imprinting in mammalian development. Cell 96:185–193

Tucker KL, Talbot D, Lee MA, Leonhardt H, Jaenisch R (1996) Complementation of methylation deficiency in embryonic stem cells by a DNA-methyltransferase minigene. Proc Natl Acad Sci USA 93:12920–12925

Uchida T, Ohashi H, Kinoshita T, Saito H, Taguchi R, Hotta T, Murate T (1998) Hypermethylation of p15(INK4B) gene in a patient with acute myelogenous leukemia evolved from paroxysmal nocturnal hemoglobinuria. Blood 92:2981–2983

Veigl ML, Kasturi L, Olechnowicz J, Ma AH, Lutterbaugh JD, Periyasamy S, Li GM, Drummond J, Modrich PL, Sedwick WD, Markowitz SD (1998) Biallelic inactivation of hMLH1 by epigenetic gene silencing, a novel mechanism causing human MSI cancers. Proc Natl Acad Sci USA 95:8698–8702

Versteege I, Sevenet N, Lange J, Rousseau-Merck MF, Ambros P, Handgretinger R, Aurias A, Delattre O (1998) Truncating mutations of hSNF5/INI1 in aggressive paediatric cancer. Nature 394:203–206

Vertino PM, Yen R-WC, Gao J, Baylin SB (1996) De Novo methylation of CpG island sequences in human fibroblasts overexpressing DNA (cytosine-5)-methyltransferase. Mol and Cell Biol 16: 4555–4565

Vleminckx K, Vakaet L, Mareel M, Fiers W, Van Roy F (1991) Genetic manipulation of E-cadherin expression by epithelial tumor cells reveals an invasion suppressor role. Cell 66:107–119

Wang W (1996) Diversity and specialisation of mammalian SWI/SNF complexes. Genes and Dev 10:2117–2130

Watts GS, Pieper RO, Costello JF, Peng Y-M, Dalton WS, Futscher BW (1997) Methylation of discrete regions of the $O^6$-methylguanine DNA methyltransferase (MGMT) CpG island is associated with heterochromatinization of the MGMT transcription start site and silencing of the gene. Mol and Cell Biol 17:5612–5619

Weinberg RA (1995) The retinoblastoma protein and cell cycle control. Cell 81:323–330

Willemze R, Archimbaud E, Muus P (1993) Preliminary results with 5-aza-2'-deoxycytidine (DAC)-containing chemotherapy in patients with relapsed or refractory acute leukemia. Leukemia 7:49–50

Wong DJ, Barrett MT, Stoger R, Emond MJ, Reid BJ (1997) p16INK4a promoter is hypermethylated at a high frequency in esophageal adenocarcinomas. Cancer Res 57:2619–2622

Wong IHN, Lo YMD, Zhang J, Liew C-T, Ng MHL, Wong N, Lai PBS, Lau WY, Hjelm NM, Johnson PJ (1999) Detection of aberrant p16 methylation in the plasma and serum of liver cancer patients. Cancer Res 59:71–73

Wu J, Issa J-P, Herman J, Bassett DE Jr, Nelkin BD, Baylin SB (1993) Expression of an exogenous eukaryotic DNA-methyltransferase gene induces transformation of NIH 3T3 cells. Proc Natl Acad Sci USA 90:8891–8895

Yen RW, Vertino PM, Nelkin BD, Yu JJ, el-Deiry W, Cumaraswamy A, Lennon GG, Trask BJ, Celano P, Baylin SB (1992) Isolation and characterization of the cDNA encoding human DNA methyltransferase. Nucleic Acids Research 20:2287–2291

Yoder JA, Yen R-WC, Vertino PM, Bestor TH, Baylin SB (1996) New 5' regions of the murine and human genes for DNA (cytosine- 5)-methyltransferase. J Biol Chem 271:31092–31097

Yoshiura K, Kanai Y, Ochiai A, Shimoyama Y, Sugimura T, Hirohashi S (1995) Silencing of the E-cadherin invasion-suppressor gene by CpG methylation in human carcinomas. Proc Natl Acad Sci USA 92:7416–7419

Zagonel V, Re GI, Marotta G, Babare R, Sardeo G, Gattei V, De Angelis V, Monfardini S, Pinto A (1993) 5-Aza-2'-deoxycytidine (decitabine) induces trilineage response in unfavourable myelodysplastic syndromes. Leukemia 7:30–35

# Mammalian Methyltransferases and Methyl-CpG-Binding Domains: Proteins Involved in DNA Methylation

B. Hendrich and A. Bird

## 1 Introduction

The modified base 5-methylcytosine has been known to exist in mammalian DNA since 1950 (Wyatt 1950). It wasn't until 1988 that the gene encoding the enzyme responsible for the maintenance of 5-methylcytosine in mammals, DNA-(cytosine-5) methyltransferase 1 (DNMT1), was identified (Bestor et al. 1988). The following year, a protein activity was reported; it was able to bind DNA containing methylated cytosine followed by guanosine (MeCpG) but, otherwise, it was indifferent to the sequence context (Meehan et al. 1989). A different protein activity, which was also capable of binding the sequence MeCpG, was identified in 1992, and the corresponding gene was cloned (Lewis et al. 1992). This provided the first molecular handle on MeCpG-binding proteins (MeCPs). For the following 5 years, however, no further proteins were identified that were able to either methylate DNA or to bind specifically to methylated DNA. The past 2 years have seen a flurry of activity in this field, with the reporting of three new candidate methyl-

Institute of Cell and Molecular Biology, University of Edinburgh, Darwin Building, King's Buildings, Edinburgh EH9 3JR, Scotland, UK
E-mail: Brian.Hendrich@ed.ac.uk
E-mail: A.Bird@ed.ac.uk

transferases (OKANO et al. 1998b; YODER and BESTOR 1998) and four new candidate MeCPs (CROSS et al. 1997; HENDRICH and BIRD 1998). Also developing is an
ever-more-precise molecular picture of exactly how DNA methylation affects
transcription (NAN et al. 1997, 1998; BESTOR 1998; JONES et al. 1998; NG et al.
1999; WADE et al. 1999; ZHANG et al. 1999). In this chapter, we will review what is
known about the mammalian proteins involved in both methylating DNA and in
interpreting the signal that DNA methylation represents. For an evolutionary
discussion of the known eukaryotic DNMTs, we refer the reader to a recent review
by Colot and Rossignol (1999). In this review, we will only discuss MeCPs that do
not require additional DNA sequence for specific binding to DNA. For a summary
of MeCPs in general we refer the reader to a review by TATE and BIRD (1993).

## 2  DNA-(Cytosine-5) Methyltransferases

### 2.1  DNA-(Cytosine-5) Methyltransferase 1

Methylation of DNA at the 5 position of cytosine occurs in organisms ranging
from mammals to bacteria. There are two types of methylation that might happen
in a cell: *de novo* methylation and maintenance methylation. When a site that was
previously unmethylated becomes methylated, the site is said to have been methylated *de novo*. When DNA containing a symmetrically methylated CpG dinucleotide is replicated, the result is two double-stranded DNA molecules, each
containing a methylated CpG dinucleotide on the parental strand but containing an
unmethylated CpG dinucleotide on the newly synthesised strand. The methylated
state of the site on the parent molecule is maintained in the daughter molecules
when a maintenance methyltransferase recognises the hemimethylated site and
methylates the unmethylated cytosine, restoring the symmetrically methylated CpG
dinucleotide pair. A universal feature of the enzymes that catalyse the methylation
reaction is the presence in each of a set of six highly conserved motifs which
comprise the catalytic domain (KUMAR et al. 1994). DNMT1, the first mammalian
DNMT to be identified, comprises the signature catalytic domain located in the
C-terminal half of the protein and an N-terminal region of approximately 1000
amino acids (Fig. 1A; Bestor et al. 1988). Functions attributed to this N-terminal
domain include conferring a specificity for hemimethylated DNA (BESTOR 1992)
and targeting the protein to the replication machinery (LEONHARDT et al. 1992;

---

**Fig. 1.** Comparison of murine DNA-(cytosine-5) methyltransferases (**A**) and methyl-CpG-binding proteins (**B**). The proteins are situated to highlight the positions of the catalytic domains (indicated in *black
boxes* labelled "Catalytic Domain") of the methyltransferases in **A** and the methyl-CpG binding domains
(indicated in a *grey box* labelled "MBD") of the methyl-CpG binding proteins in **B**. Other known
functional domains are indicated in *striped boxes*, while simple amino acid repeats are indicated in
*vertical stripes*. The scale in Part B is not the same as that in Part A

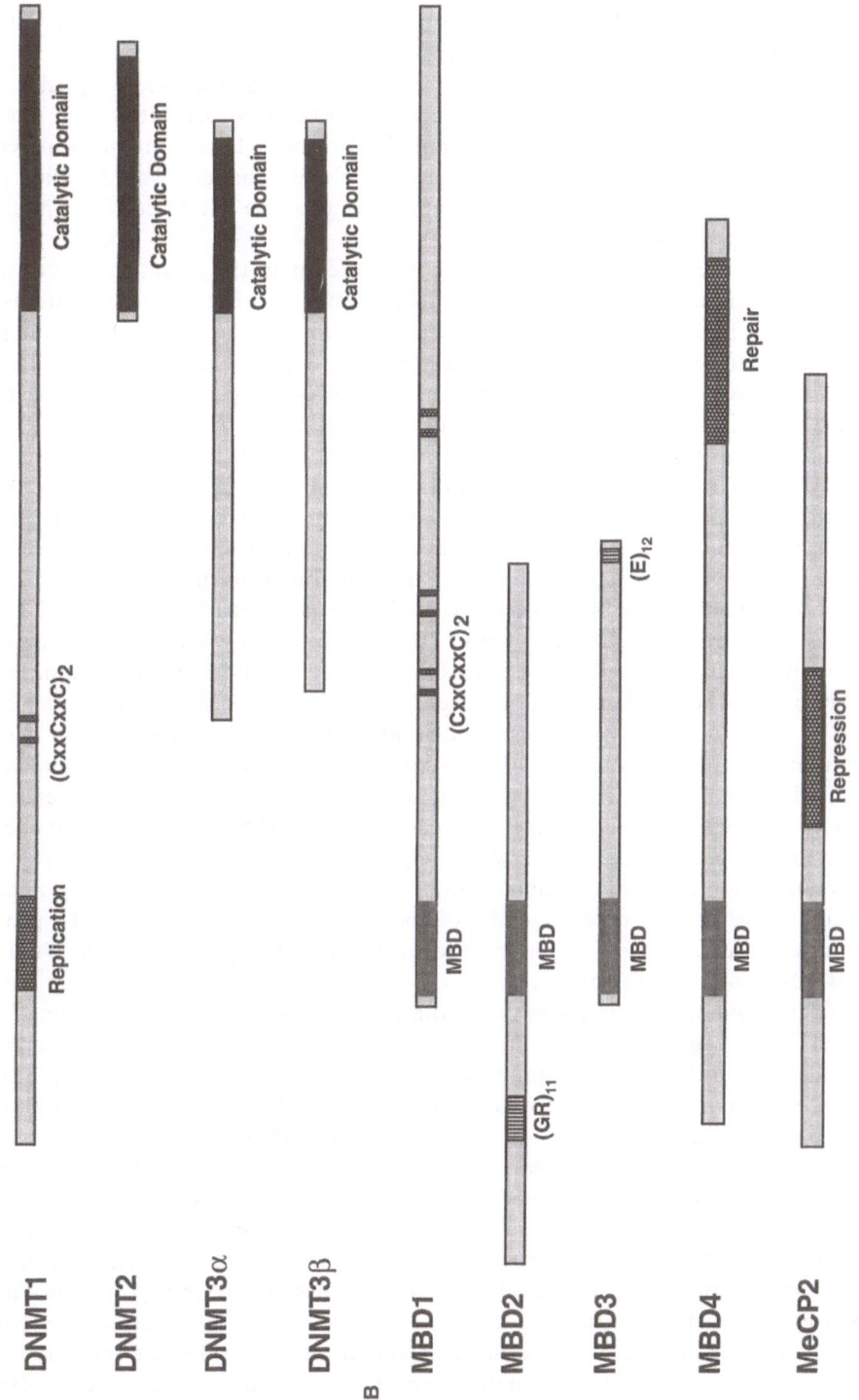

CHUANG et al. 1997). DNMT1 has the properties expected of an enzyme responsible for maintenance methylation: it is ubiquitously expressed in somatic tissues, interacts with replication machinery at replication foci and is able to quickly methylate hemimethylated DNA. Thus, because DNMT1 is associated with replication machinery, it can regenerate the symmetrically methylated state of the parental DNA molecule. This scenario contributes to the maintenance of methylation patterns in somatic cells, where the methylation status of the parent cell is maintained in subsequent daughter cells.

Maintenance methylation cannot, however, explain the establishment of methylation patterns in germ cells or in early embryos. Methylation patterns of mammalian male and female germ cells differ from each other and from the patterns found in somatic cells (MONK et al. 1987; SANFORD et al. 1987; DRISCOLL and MIGEON 1990). In addition, fluctuations in DNA-methylation levels have been reported to occur in early murine embryos. Specifically, an overall loss of methylation levels has been found to occur at numerous genomic sites in the murine embryo prior to implantation; these are then restored by a wave of de novo methylation at approximately the time of implantation (MONK et al. 1987; KAFRI et al. 1992). In order to establish the different methylation patterns that have been described, the cell needs to be able to lose methylation from fully methylated DNA (demethylation) and to symmetrically methylate previously unmethylated DNA (de novo methylation). Demethylation can be achieved by the failure of maintenance methylation in replicating cells. Good evidence exists that this is indeed the mechanism by which demethylation occurs in early embryogenesis (ROUGIER et al. 1998). De novo methylation must involve an active process distinct from that of maintenance methylation. The observation that removal of the N-terminal regulatory domain from the catalytic domain of DNMT1 results in a marked increase in de novo methylation activity led to the realisation that DNMT1 might itself encode the de novo methyltransferase activity required in embryogenesis (BESTOR 1992). This need not require an alternate form of the enzyme or modification by cofactors specific to germ cells or early embryos, because the full-length DNMT1 protein contains significant de novo methylation activity in vitro (BESTOR and VERDINE 1994; OKANO et al. 1998b).

This possibility led Bestor and colleagues to investigate the nature of the DNMT1 message and protein in mammalian germ cells and early development (CARLSON et al. 1992; MERTINEIT et al. 1998). DNMT1 activity is regulated in germ cells through the use of alternate promoters and 5' exons in the Dnmt1 gene. Despite having large amounts of Dnmt1 RNA, no DNMT1 protein can be detected in the nuclei of oocytes or early embryos until after implantation. Pachytene spermatocytes contain large amounts of Dnmt1 RNA, but the use of a pachytene-specific promoter (and first exon) results in the production of a RNA that is, apparently, untranslated (MERTINEIT et al. 1998). DNMT1 protein can be found in oocytes of postnatal ovaries, but the protein is sequestered to the cytoplasm. This is apparently due to the use of a tissue-specific promoter (and first exon), which produces a messenger RNA encoding a slightly truncated protein of about 175kDa, compared with the 190-kDa somatic form of the protein (MERTINEIT et al. 1998).

Such a truncated form of the protein might be exactly what is expected should the *Dnmt1* gene also produce a *de novo* methyltransferase. Contrary to the expectation for a *de novo* methyltransferase, the truncated DNMT1 protein persists in the cytoplasm of cells in the early embryo until after implantation. After this time, the somatic promoter and first exon are used to produce the somatic form of DNMT1 (CARLSON et al. 1992; TUCKER et al. 1996; YODER et al. 1996). Thus, there is no detectable nuclear DNMT1 protein in the cells of pre-implantation embryos except at the eight-cell stage.

DNMT1 translocates briefly into the nuclei of pre-implantation embryos at around the eight-cell stage only to be sequestered in the cytoplasm again by the blastocyst stage (CARLSON et al. 1992). These observations are compatible with the phenotype of DNMT1-deficient embryos. Murine embryos lacking a functional *Dnmt1* gene are normal during pre-implantation development but begin to show developmental delay at about 9.5 days post-coitus (d.p.c.) and fail to develop past 12.5 d.p.c. (LI et al. 1992). The two sets of results, taken together, seem to indicate that DNMT1 is not required during pre-implantation development in mice. Indeed, embryonic stem cells lacking DNMT1 grow apparently normally (LI et al. 1992; LEI et al. 1996) and are able to contribute to embryos when rescued with a *Dnmt1* complementary DNA (TUCKER et al. 1996).

The fact that DNMT1 is cytoplasmic during nearly all of pre-implantation development hints at the presence of a separate *de novo* methyltransferase which is responsible for the observed wave of DNA methylation, though it remains possible that undetectable levels of nuclear DNMT1 are present and active in pre-implantation embryos, or that the DNMT1 present in eight-cell embryos carries out *de novo* methylation. Until recently, direct proof for the existence of a *de novo* methyltransferase distinct from DNMT1 was lacking. Such proof came after careful analysis of DNMT1-deficient cells (LEI et al. 1996). Embryonic stem cells lacking a functional *Dnmt1* gene retained low but detectable levels of DNA methylation for more than 20 cell divisions. *De novo* DNA methylation was detectable after infection with a murine retrovirus. Together, these results indicate that a *de novo* methyltransferase is present in embryonic stem (ES) cells and is capable of methylating incoming parasitic elements and maintaining low levels of DNA methylation in the absence of DNMT1. Three different genes, all of which encode potential *de novo* methyltransferase candidates, have now been identified independently by two different groups (Fig. 1A; OKANO et al. 1998a, 1998b; YODER and BESTOR 1998; Bestor, personal communication).

## 2.2 DNA-(Cytosine-5) Methyltransferase 2

The first new methyltransferase candidate to be reported (YODER and BESTOR 1998) was identified by searching the expressed-sequence-tag (EST) databases for genes with the potential to encode protein motifs common to all (cytosine-5) methyltransferases (POSFAI et al. 1989). A gene which could encode a protein with all of the diagnostic methyltransferase motifs was found and was thus named DNMT2

(Fig. 1A; YODER and BESTOR 1998). Whereas DNMT1 contains a large N-terminal regulatory domain in addition to the catalytic C-terminal domain, DNMT2 consists of only the catalytic domain (YODER and BESTOR 1998). Bestor has shown that, when freed from the N-terminal regulatory domain, the catalytic domain of DNMT1 has increased rates of *de novo* methylation compared with the intact protein (BESTOR 1992). For this reason, one might predict that a *de novo* methyltransferase will consist of only a catalytic domain, so DNMT2 seems a promising candidate – or would be if it could be shown that it has (cytosine-5) methyltransferase activity. No DNA methyltransferase activity was detectable with the bacterially expressed, recombinant protein (YODER and BESTOR 1998). En Li and colleagues independently identified the DNMT2 gene and were similarly unable to demonstrate any methyltransferase activity in vitro (OKANO et al. 1998a). DNMT2 was effectively eliminated as a candidate for the *de novo* methyltransferase when Okano et al. showed that deletion of the gene in ES cells had no effect upon the *de novo* methylation activities identified in DNMT1-deficient cells (OKANO et al. 1998a).

If DNMT2 isn't a DNA methyltransferase, then what is its function? ES cells lacking DNMT2 are apparently normal, though the phenotype of DNMT2-deficient mice has not yet been reported. One possible avenue for further investigation is an organism in which 5-methylcytosine has not been detected: *Schizosaccharomyces pombe*. DNMT2 is similar to a protein identified in fission yeast called pmt1p (WILKINSON et al. 1995). Pmt1p is a methyltransferase-like protein for which, like DNMT2, no methyltransferase activity has been demonstrated. Lack of activity appears to be due to the mutation of an amino acid at what is predicted to be the catalytic site (WILKINSON et al. 1995). Restoration of this single amino acid in pmt1p to that of the consensus sequence resulted in a protein that did contain DNA-methyltransferase activity (PINARBASI et al. 1996). Although DNMT2 has the correct amino acid at the active site (YODER and BESTOR 1998), it still lacks DNA-methyltransferase activity. Yeast strains lacking pmt1p have been generated but have no obvious phenotype (WILKINSON et al. 1995) so, at the moment, the function of pmt1p remains unknown.

## 2.3 DNMT3α and DNMT3β

Two further candidates for the missing *de novo* methyltransferases are DNMT3α and DNMT3β (Fig. 1A; OKANO et al. 1998b). These two proteins were again identified by screening sequence databases for genes capable of encoding methyltransferase-like motifs. DNMT3α and DNMT3β are very similar proteins but are encoded by different genes. Both DNMT3α and DNMT3β have demonstrable *de novo* methyltransferase activity in in vitro assays (OKANO et al. 1998b). The rate at which these two proteins methylate unmethylated DNA is still less than that of DNMT1 but, whereas DNMT3α and DNMT3β show no preference between unmethylated or hemimethylated DNA as a substrate, the preferred substrate for DNMT1 is definitely hemimethylated DNA (OKANO et al. 1998b). Despite their low *de novo* methyltransferase activity, might DNMT3α and DNMT3β be the long sought-after *de novo* methyltransferases? One requirement for a *de novo* methyl-transferase is that it be expressed when *de novo* methylation is expected to happen, i.e. before implantation of mammalian embryos. Both DNMT3α and DNMT3β are strongly expressed in undifferentiated ES cells, but expression is greatly reduced on differentiation into embryoid bodies (OKANO et al. 1998b). Expression in embryoid bodies mimics that seen in somatic tissues. Thus, both genes are well expressed in an undifferentiated cell type and poorly expressed in a differentiated cell type. Whether the genes show similar expression patterns in pre- and post-implantation development remains to be seen. Given the sequestration of DNMT1 protein in early development, it will also be important to characterise the subcellular locali-sation of DNMT3α and DNMT3β proteins in pre-implantation embryos before declaring the search for the *de novo* methyltransferase over. What of the expected phenotype of a *de novo* methyltransferase knockout? Will ES cells survive without a *de novo* methyltransferase? Only time and the generation of knockouts will tell.

# 3 MeCpG-Binding Proteins

Methylation of promoter DNA in mammalian cells can, and frequently does, result in transcriptional silencing of that promoter. Until recently, the molecular mech-anisms by which this transcriptional repression was achieved were not certain. Initially, it was believed that repression would be achieved by the masking of DNA-sequence motifs (due to methylation of CpG dinucleotides) normally recognised by a sequence-specific transcription factor (Fig. 2, direct model; RAZIN and RIGGS 1980; DORFLER 1983; EDEN and CEDAR 1994). The methyl group on 5-met-hylcytosine protrudes into the major groove of the DNA and, thus, methylation of a cytosine could render a site unrecognisable to a sequence-specific DNA-binding protein. The precedent for this direct interference model is the well-known re-striction-modification system of bacteria, in which the activity of certain restriction enzymes is prevented by methylation of their target sites (ARBER and LINN 1969). In keeping with this direct model of transcriptional inhibition, a number of tran-scription factors are known to be incapable of binding to methylated versions of their target sequences (WATT and MOLLOY 1988; IGUCHI-ARIGA and SCHAFFNER 1989; COMB and GOODMAN 1990; INAMDAR et al. 1991; TATE and BIRD 1993). However, genes dependent on the activity of these methylation-sensitive tran-scription factors are not necessarily regulated by DNA methylation. Inconsistent with the direct model is the fact that several general transcription factors such as Sp1 and CCAAT-binding transcription factor have been shown to be impervious to the methylation status of their target sequences (HARRINGTON et al. 1988; HÖLLER et al. 1988; BEN-HATTAR et al. 1989). Thus, the direct model cannot explain all instances of silencing by methylation.

The observations that repression of methylated transgenes only occurs after chromatin assembly (BUSCHHAUSEN et al. 1987; KASS et al. 1997), that DNA

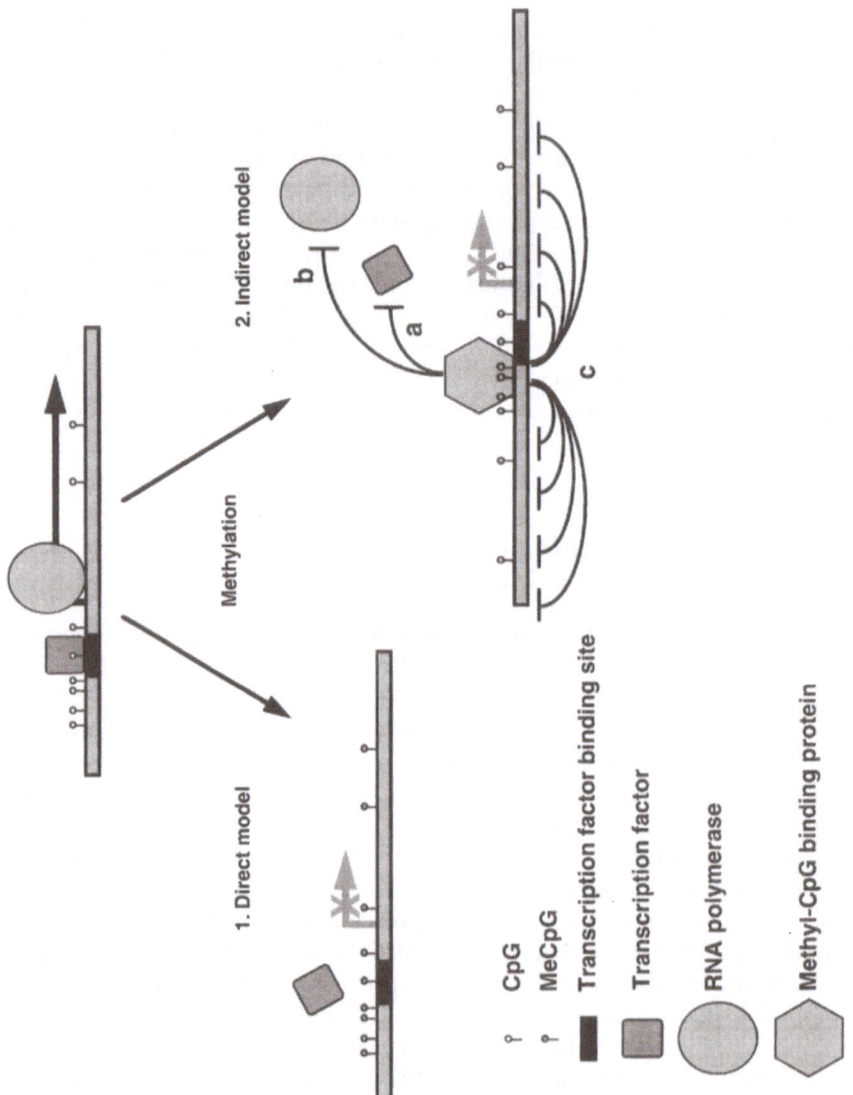

methylation is capable of repressing transcription at some distance (KASS et al. 1993; HSIEH 1997) and that fully methylated genes can be transcribed in some circumstances (BOYES and BIRD 1991; LEVINE et al. 1991; KASS et al. 1997) indicate that methylation-mediated repression is primarily achieved through a more indirect mechanism (Fig. 2, indirect model; TATE and BIRD 1993). Under the indirect model, the sequence MeCpG is recognised by a MeCP, which then prevents transcription from a methylated promoter (Fig. 2). Transcription inhibition might be brought about by: (1) steric hindrance, in which the binding of the MeCP blocks access of the transcription factor to the DNA; (2) interference, in which the presence of the MeCP prevents loading or activity of the RNA polymerase; or (3) chromatin modulation, in which the MeCP invokes a change in the local chromatin structure from a transcriptionally competent to a transcriptionally incompetent state (Fig. 2).

## 3.1 MeCpG-Binding Protein 1

Two protein activities, called MeCP1 and MeCP2, have been identified; both bind to methylated DNA irrespective of sequence context and are capable of repressing transcription (MEEHAN et al. 1989; BOYES and BIRD 1991; LEWIS et al. 1992; NAN et al. 1997), making them likely candidates for mediators of indirect silencing. MeCP1 is a large protein complex which requires more than ten MeCpGs to bind strongly to methylated DNA (MEEHAN et al. 1989). Thus, MeCP1 may only be able to bind to regions of the genome in which the density of methylation is high. Consistent with this hypothesis, Boyes and Bird (and later Hsieh) found that the ability of MeCP1 to repress transcription from a methylated template depends on the density of methylation present in that sequence (BOYES and BIRD 1992; HSIEH 1994). These observations are consistent with a scenario in which MeCP1 is responsible for repressing genes when the promoter DNA contains multiple MeCpGs. Around 60% of RNA-polymerase-II-transcribed genes contain promoters located within CpG islands (ANTEQUERA and BIRD 1994) and, thus, would be very dense in the methylatable sequence CpG. However, CpG islands are, for the most part, unmethylated. CpG-island methylation does occur on the inactive X chromosome of female eutherian mammals (RIGGS and PFEIFER 1992), at imprinted genes (RAZIN and CEDAR 1994), and in some tumours (DE BUSTROS et al. 1988;

Fig. 2. Two potential models of DNA-methylation-mediated transcriptional repression. The *top panel* shows an unmethylated promoter on which a sequence-specific transcription factor (*Rectangle*) binds to its recognition sequence (*black box*) and allows for transcription of the gene by RNA polymerase II (*circle*). Upon methylation of the promoter, transcription is inhibited. In the direct model, methylation of a single cytosine residue makes the sequence unrecognisable to the transcription factor and, thus, transcription is prevented. There are three possible scenarios for the indirect model, all involving the binding of a methyl-CpG-binding protein (MeCP, *hexagon*). In scenario *a*, the binding of the MeCP covers the transcription-factor-binding site, making it inaccessible to binding by the transcription factor. In scenario *b*, the MeCP directly interacts with one or more components of the transcription machinery, preventing the formation of an active transcriptional complex. In scenario *c*, the binding of the MeCP causes a change in chromatin structure, which makes the promoter transcriptionally incompetent

SAKAI et al. 1991; HERMAN et al. 1995). DNA methylation is known to have a direct effect on transcription in X-chromosome inactivation and genomic imprinting (SHAPIRO and MOHANDAS 1982; LI et al. 1993; PANNING and JAENISCH 1996), and it might well be MeCP1 that causes transcriptional silencing of methylated promoters in these two situations. Despite years of effort, MeCP1 has proven resistant to cloning via the standard biochemical purification techniques, so the molecular nature of the MeCP1 complex remains unknown. Recently, however, molecular candidates for MeCP1 components have been identified (CROSS et al. 1997; HENDRICH and BIRD 1998; see below). One of these candidates has recently been shown to be a component of MeCP1, providing the first real insight into the molecular nature of this protein complex (NG et al. 1999; see below).

## 3.2 MeCpG-Binding Protein 2

Like MeCP1, MeCP2 was identified as a protein activity capable of binding to methylated DNA irrespective of sequence context (LEWIS et al. 1992). Unlike MeCP1, however, MeCP2 requires only a single, symmetrically methylated CpG to bind DNA (LEWIS et al. 1992). MeCP2 consists of a single polypeptide that binds methylated DNA via a distinct MeCpG-binding domain (MBD; LEWIS et al. 1992; NAN et al. 1993). The first clue to the mechanism of action of MeCP2 came when a discrete portion of the MeCP2 protein was found to be capable of repressing transcription from a reporter gene, and did so at some distance (NAN et al. 1997). This indicates that MeCP2 is able to create a domain of repression that extends a long way from the site at which it is actually bound to the DNA (Fig. 2, indirect model c). How might such a sphere of influence be achieved? Methylated DNA is known to be somewhat protected from the action of nucleases (KESHET et al. 1986; ANTEQUERA et al. 1989), indicating that it adopts a chromatin structure that is different from that of unmethylated DNA. This finding, combined with the observation that chromatin assembly is required for transcriptional silencing (BUSCHHAUSEN et al. 1987; KASS et al. 1997), led Nan et al. to investigate the possibility that the "sphere of influence", nucleated by MeCP2, that leads to transcriptional repression is manifest as a modulation of the surrounding chromatin structure (NAN et al. 1998).

The best understood method of chromatin modulation is via the acetylation or deacetylation of the core histones in the nucleosome (TURNER 1991; GRUNSTEIN 1997). Nucleosomes that contain histones with acetylated N-terminal tails are primarily found in transcriptionally competent chromatin. A possible mode of action for MeCP2 is that it recruits a histone deacetylase activity in methylated DNA, leading to the deacetylation of histone tails in the surrounding chromatin, thus establishing a "repressive chromatin domain" (KASS et al. 1997). In order to test this hypothesis, NAN et al. (1998) looked for, and found, an interaction between MeCP2 and a co-repressor complex containing histone deacetylases and the homologue of a known yeast repressor protein, mSin3 (GRUNSTEIN 1997; PAZIN and KADONAGA 1997). Not only was MeCP2 found to interact with mSin3A in immunoprecipitation experiments, the repression activity of MeCP2 was found to be sensitive to the action of trichostatin A, a drug that inhibits the activity of histone deacetylases (NAN et al. 1998). Jones et al., working with the *Xenopus laevis* MeCP2 orthologue, also demonstrated an interaction between MeCP2 and Sin3 and were able to immunoprecipitate histone deacetylase activity using anti-MeCP2 antibodies (JONES et al. 1998). Thus, it is clear that MeCP2-mediated transcriptional silencing is achieved through the deacetylation of histones (or other proteins; PAZIN and KADONAGA 1997) in chromatin.

What remains unknown is the identity of the genes normally silenced by MeCP2. Like DNMT1 (LI et al. 1992), MeCP2 is required for murine embryonic development (TATE et al. 1996); however, in contrast to DNMT1, MeCP2 is not required for somatic cell viability in vitro (Hendrich and Tate, unpublished). Furthermore, genes known to be mis-expressed in the absence of DNMT1, such as imprinted genes and the *Xist* gene (LI et al. 1993; BEARD et al. 1995), were found to be expressed normally in MeCP2-deficient somatic cell lines (Hendrich, unpublished). The fact that the two-base-pair MeCP2 DNA-binding site occurs once every 150 bases in vertebrate genomes has led to the suggestion that MeCP2 might function to prevent spurious transcription throughout the genome (BIRD 1995; NAN et al. 1997). Thus, the embryonic lethality resulting from MeCP2 deficiency could be due to low levels of inappropriate expression from numerous promoters rather than high levels of overexpression from a few target genes.

## 3.3 MBD1, MBD2, MBD3 and MBD4

The observations listed above indicate that the effects of DNA methylation are being mediated through protein factors other than just MeCP2, with MeCP1 being a prime candidate (see above). MeCP1 has proven to be remarkably resistant to purification by conventional biochemical methods and, for years after its initial publication, the molecular composition of MeCP1 remained a mystery. By 1996, however, the vast amount of EST sequence data produced by the human-genome project (BOGUSKI et al. 1993) allowed the assault on MeCP1 to be moved from the cold room to the computer room. Searching a conceptual translation of the EST databases for the amino acid sequence similar to that of the MBD of MeCP2 provided the first candidates for components of MeCP1. These searches revealed the presence, in both humans and mice, of four novel genes, each with the capacity to encode proteins containing an MBD-like motif (Fig. 1B; CROSS et al. 1997; HENDRICH and BIRD 1998). Of these four proteins (named MBD1–MBD4), three were shown to be capable of specifically binding methylated DNA both in vitro and in vivo, independent of sequence context (CROSS et al. 1997; HENDRICH and BIRD 1998). This is illustrated in Fig. 3, where green-fluorescent-protein (GFP) fusions of the MBD proteins were overexpressed in murine cells (*top panels*) or human cells (*bottom panels*). Approximately 40% of 5-methylcytosine in murine cells is found within the pericentromeric major satellite, which is visualised as 4',6-diamidino-2-phenylindole (DAPI)-bright regions in interphase nuclei (MILLER et al. 1974).

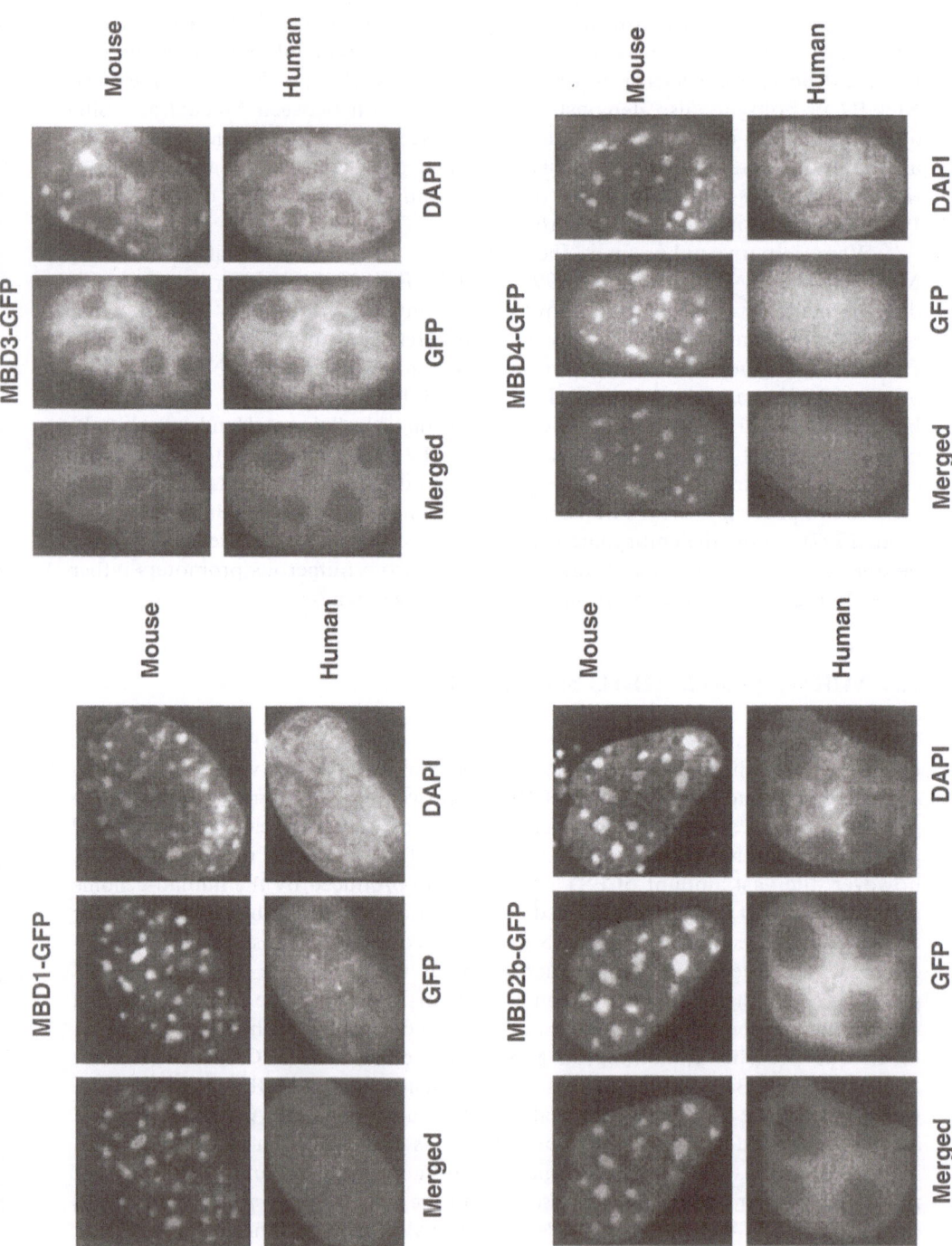

Recombinant MBD1–GFP, MBD2–GFP, and MBD4–GFP co-localise with these DAPI-bright foci, indicating a co-localisation with DNA sequences known to be enriched in 5-methylcytosine. The MBD–GFP proteins show no such localisation in human cells (which lack such high levels of MeCpG-rich satellite), demonstrating that the MBD–GFP fusion proteins are not simply binding heterochromatin. Only MBD3 showed no strict specificity for binding to methylated DNA either in band-shift experiments, southwestern assays or murine cells (Fig. 3; HENDRICH and BIRD 1998). This result was particularly surprising, because MBD3 is 71% identical to the MBD2 protein at the amino acid level, and MBD2 readily binds methylated DNA in all three assays.

MBD2 has been reported to exhibit demethylase activity (BHATTACHARYA et al. 1999), though this result has not been upheld in more recent reports (NG et al. 1999; WADE et al. 1999). Rather, both MBD2 and MBD3 have been found to be components of transcriptional repressor complexes and, like MeCP2, to associate with histone deacetylases (NG et al. 1999; WADE et al. 1999; ZHANG et al. 1999). MBD3 was found to be a component of the NuRD complex in both human cells (ZHANG et al. 1999) and *Xenopus* (WADE et al. 1999). The NuRD (or Mi-2) complex is a multi-subunit complex and has been shown to contain both chromatin-remodelling and histone deacetylase activities (WADE et al. 1998; XUE et al. 1998; ZHANG et al. 1998). Whether MBD3 directs this protein complex to methylated DNA in vivo remains somewhat controversial, as two different groups report conflicting results on this issue (WADE et al. 1999; ZHANG et al. 1999). MBD2 was shown to be a transcriptional repressor that is a component of a different histone-deacetylase-containing complex, and this protein complex was shown to be MeCP1. This finding identifies both a methylated-DNA-binding component (MBD2) and transcriptional repression components (histone deacetylases) for MeCP1. Thus, MBD2, MBD3, and MeCP2 associate with protein complexes containing histone-deacetylase activity, but all three appear to interact with different complexes and, thus, may have different in vivo functions.

Sequence analysis of the MBD1 protein revealed the presence of either two or three copies of a cysteine-rich motif (Fig. 1B), depending on alternate splicing, which is also present in DNMT1 (Fig. 1A) and the acute lymphoblastic leukaemia-1/human trithorax protein (MA et al. 1993; CROSS et al. 1997; HENDRICH and BIRD 1998). The function of this sequence motif is not known for any of these proteins (though, in the DNMT1 protein, it is capable of binding zinc; BESTOR 1992). With the exception of the MBD, the sequence of MBD1 provided no clues as to its function. Like MeCP2 and MBD2, though, MBD1 was found to be capable of

**Fig. 3.** The co-localisation of methyl-CpG-binding proteins to methylated DNA in vivo. Green fluorescent protein (GFP) fusions of methyl-CpG-binding domain (MBD)1, MBD2b, MBD3, and MBD4 were ectopically expressed in a murine fibroblast cell line (L cells, *top panels*) or a human fibrosarcoma cell line (HT1080, *bottom panels*). Of the methylcytosine in murine cells, 40% is located within major satellite DNA, which is visualised by 4′,6-diamidino-2-phenylindole staining as bright foci. MBD1–GFP, MBD2a–GFP and MBD4–GFP fusion proteins co-localise with major satellite, whereas MBD3–GFP shows diffuse nuclear staining in murine cells. All four fusion proteins show diffuse staining in the nuclei of human cells, which do not contain major satellite DNA

repressing transcription from a methylated template and, thus, represents another methylation-dependent transcriptional repressor (CROSS et al. 1997).

Searches of the protein databases with the MBD4-protein sequence produced a series of low-scoring but significant similarities to bacterial DNA-repair enzymes (HENDRICH and BIRD 1998). In particular, sequence homology indicated that MBD4 might be a DNA-mismatch glycosylase (HENDRICH et al. 1999). As was predicted by the sequence homology, MBD4 was found to contain glycosylase activity, being capable of removing a thymine or uracil when paired with guanine in a CpG context (HENDRICH et al. 1999). MBD4 was able to catalyse this reaction equally well on methylated or unmethylated templates. Specificity of the reaction for methylated DNA might be achieved through the action of the MBD-like region. The N-terminal regions of both the human and murine proteins were found to only bind to either symmetrically methylated DNA (HENDRICH and BIRD 1998) or to DNA containing a G/T or G/U mismatch in a MeCpG context. The preferred site was a G/T mismatch in the context of MeCpG (HENDRICH et al. 1999).

Homology between MBD4 and bacterial DNA-repair enzymes immediately suggests a link between DNA repair and DNA methylation in mammals. The best-known link between DNA methylation and DNA repair is the MutH/MutL/MutS repair pathway of *Escherichia coli*, in which DNA methylation is used to direct repair of base mismatches to the newly synthesised DNA strand (MODRICH 1991). Though eukaryotes are known to have a mismatch-repair system containing enzymes related to the components of the bacterial system, no methylation-dependence has yet been demonstrated for eukaryotic mismatch repair (KOLODNER 1996; MODRICH and LAHUE 1996). In particular, no homologue for MutH, the component of the bacterial system which recognises hemimethylated DNA, has been identified in eukaryotes. The possibility that, after replication, mammalian DNA remains hemimethylated long enough to invoke any such methylation-dependent DNA repair seems unlikely due to the observations that DNMT1 is targeted to DNA-replication foci (LEONHARDT et al. 1992) and that it associates directly with a component of the replication machinery (CHUANG et al. 1997). Unlike the situation in *E. coli* where DNA methylation occurs well after DNA replication (KOLODNER 1996; MODRICH and LAHUE 1996), there is no lag between replication and methylation of DNA in mammals. Given the specificity of MBD4 binding to DNA and of the glycosylase reaction itself, it is likely that the function of MBD4 is to combat the mutability of 5-methylcytosine in the genome (BOORSTEIN et al. 1989; ZUO et al. 1995; HENDRICH et al. 1999).

The mutational load that 5-methylcytosine places on the genome is not trivial. 5-Methylcytosine was first found to be a mutational hotspot in *E. coli* (COULONDRE et al. 1978). Deamination of methylcytosine produces thymine and results in conversion of a MeC/G base pair to a mismatched T/G base pair. Deamination of unmethylated cytosine produces uracil, and the resulting U/G mismatch is repaired an order of magnitude more efficiently than is a T/G mismatch, though both are repaired in favour of the G residue (LINDAHL 1982; BROWN and JIRICNY 1987). Mutation of the dinucleotide CpG to TpG or CpA occurs at a rate that is approximately 12 times higher than that of other transitions (BULMER 1986; SVED and

**Table 1.** Known properties of mammalian DNA (cytosine-5) methyltransferases (DNMTs)

|          | Maintenance methylation | De novo methylation | Required in somatic cells? | Required in ES cells? |
|----------|-------------------------|---------------------|----------------------------|-----------------------|
| DNMT1    | + [a]                   | + [b]               | +                          | −                     |
| DNMT2    | −                       | −                   | ?                          | −                     |
| DNMT3α   | −                       | + [b]               | ?                          | ?                     |
| DNMT3β   | −                       | + [b]               | ?                          | ?                     |

*ES*, embryonic stem.
[a] Demonstrated in vitro and in vivo.
[b] Demonstrated in vitro only.

**Table 2.** Known properties of methyl-CpG-binding proteins

|       | In vitro DNA binding[a] | In vivo DNA binding[b] | Function | HD associated? |
|-------|-------------------------|------------------------|----------|----------------|
| MeCP2 | −MG−, −GM−              | Methylated DNA         | Transcriptional repressor[c] | + [c] |
| MBD1  | −MG−, −GM−              | Methylated DNA         | Transcriptional repressor[c] | ? |
| MBD2  | −MG−, −GM−              | Methylated DNA         | Transcriptional repressor[c] | + [c] |
| MBD3  | Nonspecific             | Nonspecific            | ?        | + [c] |
| MBD4  | −MG−, −GT−              | Methylated DNA         | G/T mismatch glycosylase[c] | − |

*GFP*, green fluorescent protein; *HD*, histone deacetylase; *MBD*, MeCpG-binding domain.
[a] Preferred binding activity, as assayed by bandshifts.
[b] As assayed by transient transfections of GFP-fusion proteins, e.g. Fig. 3.
[c] Demonstrated in vitro only.

BIRD 1990), which can explain the genome-wide deficiency of CpG in vertebrates (BIRD 1980). Approximately 35% of the point mutations associated with human genetic disease are derived from deamination of 5-methylcytosine (COOPER and KRAWCZAK 1990). Thus, any protein that acts to maintain the integrity of 5-methylcytosine is expected to be important in the prevention of spontaneous mutation. How important such proteins are remains an open question.

# 4 Conclusions

At the beginning of 1997, the number of cloned DNMTs and MeCPs was one each. Over a period of about 18 months, that number has increased to four and five each, respectively, and we have only recently begun to determine what each of these proteins might be doing in the cell (Fig. 1, Tables 1, 2). While the maintenance methyltransferase activity of DNMT1 has been well characterised (BESTOR and VERDINE 1994), the fact that this protein is also capable of *de novo* methyltransferase activity is often overlooked. The fact that DNMT1 is not detectable in the nuclei of female germ cells or early embryos may mean that some other protein is responsible for *de novo* methylation in these cell types. At present, however, it is not known whether the observed *de novo* methylation is catalysed by DNMT1 or one of the other putative methyltransferases. If DNMT1 is the *de novo* methyltransferase, then what are the in vivo functions of DNMT2, DNMT3α and DNMT3β? Murine knockouts of the newer putative methyltransferases are urgently needed to address this question.

A similar situation exists for the MeCPs. MeCP2 is the best known and most studied of these proteins, and the molecular mechanism by which it silences transcription is now understood (BESTOR 1998; JONES et al. 1998; NAN et al. 1998). What genes does it repress? We know that it is not required for the transcriptional silencing of imprinted genes or *Xist* in murine cell lines, and it has been suggested that its role is in reduction of transcriptional noise (NAN et al. 1997). MeCP1 is likely to be the mediator of transcriptional silencing in the rare situations when CpG islands do get methylated. Thus, it will be of interest to see how the absence of MBD2 affects imprinting and X-chromosome inactivation. Also, do MBD1 and/or MBD3 contribute to methyl-dependent transcriptional silencing? If so, in what way do they contribute? Whereas loss of one methyltransferase was sufficient to deregulate imprinted-gene expression (LI et al. 1993; BEARD et al. 1995; PANNING and JAENISCH 1996), it may be that the MBD proteins comprise a redundant system in which the loss of more than one protein will be required to see a failure of silencing from a methylated island.

What role might MeCPs and/or DNMTs play in cancer? There can, as yet, be no certain answer to this question, though evidence implicating DNA methylation in cancer is mounting (BAYLIN 1997; Jones and LAIRD 1999). MeCP2, MBD1 and MBD2 are all capable of repressing transcription from methylated promoters and, thus, any one or a combination of these proteins might be required for the transcriptional silencing of methylated tumour-suppressor genes, which will be described in subsequent chapters. Should it be proven that methylation is a cause of gene silencing and subsequent transformation rather than a consequence of silencing brought on by some unknown event, then it may well be that one or more methyltransferases and one or more MBD proteins are required for tumourigenesis in these situations. Indeed, it has been reported that cell transformation by *fos* requires both DNA methyltransferase and histone deacetylases (BAKIN and CURRAN 1999). Both MeCP2 and MBD2 are known to bind methylated DNA and induce transcriptional silencing via the recruitment of histone deacetylases (JONES et al. 1998; NAN et al. 1998, NG et al. 1999). Thus, hypermethylation by DNMT1 at crucial loci could lead to inappropriate binding of MeCP1 and/or MeCP2, which, in turn, would recruit histone deacetylase to induce transcriptional silencing. Without evidence that resolves the "cause or consequence" question, such a scenario remains speculative.

Mutation of DNA-repair genes has been shown to lead to predisposition to a variety of cancers (BOYER et al. 1995; MODRICH and LAHUE 1996; JIRICNY 1998). Thus, of all the methyltransferases or MeCPs known thus far, the one that seems most likely to be directly involved in cancer is MBD4, which is implicated in DNA repair rather than in gene silencing. At the present time, it is not possible to point to a situation in which deletion or overexpression of a DNMT or a MeCP leads directly to human cancer. This is a field of intense study, however, so the previous sentence may not be true for much longer.

# References

Antequera F, Bird A (1994) Number of CpG islands and genes in human and mouse. Proc Natl Acad Sci USA 90:11995–11999

Antequera F, Macleod D, Bird AP (1989) Specific protection of methylated CpGs in mammalian nuclei. Cell 58:509–517

Arber W, Linn S (1969) DNA modification and restriction. Annu Rev Biochem 38:467–500

Bakin VA, Curran T (1999) Role of DNA 5-methylcytosine transferase in cell transformation by fos. Science 283:387–390

Baylin SB (1997) Tying it all together: epigenetics, genetics, cell cycle and cancer. Science 277:1948–1949

Beard C, Li E, Jaenisch R (1995) Loss of methylation activates *Xist* in somatic but not embryonic cells. Genes Dev 9:2325–2334

Ben-Hattar J, Beard P, Jiricny J (1989) Cytosine methylation in CTF and Sp1 recognition sites of an HSV *tk* promoter: effects on transcription in vivo and on factor binding in vivo. Nucleic Acids Res 17:10179–10190

Bestor TH (1992) Activation of mammalian DNA methyltransferase by cleavage of a Zn-binding regulatory element. EMBO J 11:2611–2617

Bestor TH (1998) Methylation meets acetylation. Nature 393:311–312

Bestor TH, Verdine GL (1994) DNA methyltransferases. Curr Op Cell Biol 6:380–389

Bestor TH, Laudano AP, Mattaliano R, Ingram VM (1988) Cloning and sequencing of a cDNA encoding DNA methyltransferase of mouse cells. The carboxyl-terminal domain of the mammalian enzymes is related to bacterial restriction methyltransferases. J Mol Biol 203:971–983

Bhattacharya SK, Ramchandani S, Cervoni N, Szyf M (1999) A mammalian protein with specific de-methylase activity for mCpG DNA. Nature 397:579–583

Bird AP (1980) DNA methylation and the frequency of CpG in animal DNA. Nucl Acids Res 8:1499–1504

Bird AP (1995) Gene number, noise reduction and biological complexity. Trends Genet 11:94–100

Boguski MS, Lowe TM, Tolstoshev CM (1993) dbEST-database for expressed sequence tags. Nat Genet 4:332–333

Boorstein RJ, Chiu LN, Teebor GW (1989) Phylogenetic evidence of a role for 5-hydroxymethyluracil-DNA glycosylase in the maintenance of 5-methylcytosine in DNA. Nucl Acids Res 17:7653–7661

Boyer JC, Umar A, Risinger JI, Lipford JR, Kane M, Yin S, Barrett JC, Kolodner RD, Kunkel TA (1995) Microsatellite instability, mismatch repair deficiency and genetic defects in human cancer cell lines. Cancer Res 55:6063–6070

Boyes J, Bird A (1991) DNA methylation inhibits transcription indirectly via a methyl-CpG binding protein. Cell 64:1123–1134

Boyes J, Bird A (1992) Repression of genes by DNA methylation depends upon CpG density and promoter strength: evidence for involvement of a methyl-CpG binding protein. EMBO J 11:327–333

Brown TC, Jiricny J (1987) A specific mismatch repair event protects mammalian cells from loss of 5-methylcytosine. Cell 50:945–950

Bulmer M (1986) Neighboring base effects on substitution rates in pseudogenes. Mol Biol Evol 3:322–329

Buschhausen G, Wittig B, Graessmann M, Graessmann A (1987) Chromatin structure is required to block transcription of the methylated herpes simplex virus thymidine kinase gene. Proc Natl Acad Sci USA 84:1177–1181

Carlson LL, Page AW, Bestor TH (1992) Properties and localization of DNA methyltransferase in pre-implantation mouse embryos: implications for genomic imprinting. Genes Dev 6:2536–2541

Chuang LS-H, Ian H-I, Koh T-W, Ng H-H, Xu G, Li BFL (1997) Human DNA-(cytosine-5) methyl-transferase–PCNA complex as a target for p21$^{WAF1}$. Science 277:1996–2000

Colot V, Rossignol J-L (1999) Eukaryotic DNA methylation as an evolutionary device. Bioessays 21: 402–11

Comb M, Goodman HM (1990) CpG methylation inhibits proenkephalin gene expression and binding of the transcription factor AP-2. Nucl Acids Res 18:3975–3982

Cooper DN, Krawczak M (1990) The mutational spectrum of single base-pair substitutions causing human genetic disease: patterns and predictions. Hum Genet 85:55–74

Coulondre C, Miller JH, Farabough PJ, Gilbert W (1978) Molecular basis of base substitution hotspots in *Escherichia coli*. Nature 274:775–780

Cross S, Meehan R, Nan X, Bird A (1997) A component of the transcriptional repressor MeCP1 shares a motif with DNA methyltransferase and HRX proteins. Nat Genet 13:256–259

De Bustros A, Nelkin BD, Silverman A, Ehrlich G, Poiesz B, Baylin SB (1988) The short arm of chromosome 11 is a "hot spot" for hypermethylation in human neoplasia. Proc Natl Acad Sci USA 85:5693–5697

Dorfler W (1983) DNA methylation and gene activity. Annu Rev Biochem 52:93–124

Driscoll D, Migeon B (1990) Sex differences in methylation of single-copy genes in human meiotc germ cells: implications for X chromosome inactivation, parental imprinting, and origin of CpG mutations. Som Cell Mol Genet 16:267–282

Eden S, Cedar H (1994) Role of DNA methylation in the regulation of transcription. Curr Op Genet Dev 4:255–259

Grunstein M (1997) Histone acetylation in chromatin structure and transcription. Nature 389:349–352

Harrington MA, Jones PA, Imagawa M, Karin M (1988) Cytosine methylation does not affect binding of transcription factor Sp1. Proc Natl Acad Sci USA 85:2066–2070

Hendrich B, Bird A (1998) Identification and characterization of a family of mammalian methyl-CpG-binding proteins. Mol Cell Biol 18:6538–6547

Hendrich B, Hardeland U, Ng H-H, Jiricny J, Bird A (1999) The thymine glycosylase MBD4 can bind to the product of deamination at methylated CpG sites. Nature 401:301–304

Herman JG, Merlo A, Mao L, Lapidus RG, Issa JP, Davidson NE, Sidransky D, Baylin S (1995) Silencing of the VHL tumor-suppressor gene by DNA methylation in renal carcinoma. Proc Natl Acad Sci USA 91:9700–9704

Höller M, Westin G, Jiricny J, Schaffner W (1988) Sp1 transcription factor binds DNA and activates transcription even when the binding site is CpG methylated. Genes Dev 2:1127–1135

Hsieh C-L (1994) Dependence of transcriptional repression on CpG methylation density. Mol Cell Biol 14:5487–5494

Hsieh C-L (1997) Stability of patch methylation and its impact in regions of transcriptional initiation and elongation. Mol Cell Biol 17:5897–5904

Iguchi-Ariga SMM, Schaffner W (1989) CpG methylation of the cAMP responsive enhancer/promoter sequence TGACGTCA abolishes specific factor binding as well as transcriptional activation. Genes Dev 3:612–619

Inamdar NM, Ehrlich KC, Ehrlich M (1991) CpG methylation inhibits binding of several sequence-specific DNA-binding proteins from pea, wheat, soybean and cauliflower. Plant Molecular Biology 17:111–123

Jiricny J (1998) Eukaryotic mismatch repair: an update. Mutat Res 409:107–121

Jones PA, Laird PW (1999) Cancer epigenetics comes of age. Nature Genetics 21:163–167

Jones PL, Veenstra GJC, Wade PA, Vermaak D, Kass SU, Landsberger N, Strouboulis J, Wolffe AP (1998) Methylated DNA and MeCP2 recruit histone deacetylase to repress transcription. Nat Genet 19:187–191

Kafri T, Ariel M, Brandeis M, Shemer R, Urven L, McCarrey J, Cedar H. Razin A (1992) Developmental pattern of gene-specific DNA methylation in the mouse embryo and germ line. Genes Dev 6:705–714

Kass SU, Goddard JP, Adams RLP (1993) Inactive chromatin spreads from a focus of methylation. Mol Cell Biol 13:7372–7379

Kass SU, Landsberger N, Wolffe AP (1997a) DNA methylation directs a time-dependent repression of transcription initiation. Curr Biol 7:1570165

Kass SU, Pruss D, Wolffe AP (1997b) How does DNA methylation repress transcription? Trends Genet 13:444–449

Keshet I, Leiman-Hurwitz J, Cedar H (1986) DNA methylation affects the formation of active chromatin. Cell 44:535 543

Kolodner R (1996) Biochemistry and genetics of eukaryotic mismatch repair. Genes Dev 10:1433–1442

Kumar S, Cheng X, Klimasauskas S, Mi S, Posfai J, Roberts RJ, Wilson GG (1994) The DNA-(cytosine-5) methyltransferases. Nucl Acids Res 22:1–10

Lei H, Oh SP, Okano M, Jüttermann R, Goss KA, Jaenisch R, Li E (1996) De novo DNA cytosine methyltransferase activities in mouse embryonic stem cells. Development 122:3195–3205

Leonhardt H, Page AW, Weier H-U, Bestor TH (1992) A targeting sequence directs DNA methyltransferase to sites of DNA replication in mammalian nuclei. Cell 71:865–873

Levine A, Cantoni GL, Razin A (1991) Inhibition of promoter activity by methylation: possible involvement of protein mediators. Proc Natl Acad Sci USA 88:6515–6518

Lewis JD, Meehan RR, Henzel WJ, Maurer-Fogy I, Jeppesen P, Klein F, Bird A (1992) Purification, sequence, and cellular localization of a novel chromosomal protein that binds to methylated DNA. Cell 69:905–914

Li E, Bestor TH, Jaenisch R (1992) Targeted mutation of the DNA methyltransferase gene results in embryonic lethality. Cell 69:915–926

Li E, Beard C, Jaenisch R (1993) Role for DNA methylation in genomic imprinting. Nature 366:362–365

Lindahl T (1982) DNA repair enzymes. Annu Rev Biochem 5:61–87

Ma Q, Alder H, Nelson KK, Chatterjee D, Gu Y, Nakamura T, Canaani E, Croce CM, Siracusa LD, Buchberg AM (1993) Analysis of the murine All-1 gene reveals conserved domains with human ALL-1 and identifies a motif shared with DNA methyltransferases. Proc Natl Acad Sci USA 90:6350–6354

Meehan RR, Lewis JD, McKay S, Kleiner EL, Bird AP (1989) Identification of a mammalian protein that binds specifically to DNA containing methylated CpGs. Cell 58:499–507

Mertineit C, Yoder JA, Taketo T, Laird DW, Trasler JM, Bestor TH (1998) Sex-specific exons control DNA methyltransferase in mammalian germ cells. Development 125:889–897

Miller OL, Schnedl W, Allen J, Erlanger BF (1974) Immunofluorescent localization of 5MC. Nature 251:636–637

Modrich P (1991) Mechanisms and biological effects of mismatch repair. Ann Rev Genet 25:229–253

Modrich P, Lahue R (1996) Mismatch repair in replication fidelity, genetic recombination, and cancer biology. Ann Rev Biochem 65:101–133

Monk M, Boubelik M, Lehnert S (1987) Temporal and regional changes in DNA methylation in the embryonic, extraembryonic and germ cell lineages during mouse embryo development. Development 99:371–382

Nan X, Meehan RR, Bird A (1993) Dissection of the methyl-CpG binding domain from the chromosomal protein MeCP2. Nucleic Acids Res 21:4886–4892

Nan X, Campoy F, Bird A (1997) MeCP2 is a transcriptional repressor with abundant binding sites in genomic chromatin. Cell 88:471–481

Nan X, Ng H-H, Johnson CA, Laherty CD, Turner BM, Eisenmann RN, Bird A (1998) Transcriptional repression by the methyl-CpG-binding protein MeCP2 involves a histone deacetylase complex. Nature 393:386–389

Ng H-H, Zhang Y, Hendrich B, Johnson CA, Turner BM, Erdjument-Bromage H, Tempst P, Reinberg D, Bird A (1999) MBD2 is a transcriptional repressor belonging to the MeCP1–histone-deacetylase complex. Nature Genetics 23:58–61

Okano M, Xie S, Li E (1998a) Dnmt2 is not required for de novo and maintenance methylation of viral DNA in embryonic stem cells. Nucl Acids Res 26:2536–2540

Okano M, Xie S, Li E (1998b) Cloning and characterization of a family of novel mammalian DNA-(cytosine-5) methyltransferases. Nature Genetics 19:219–220

Panning B, Jaenisch R (1996) DNA hypomethylation can activate Xist expression and silence X-linked genes. Genes Dev 10:1991–2002

Pazin MJ, Kadonaga JT (1997) What's up and down with histone deacetylation and transcription? Cell 89:325–328

Pinarbasi E, Elliott J, Hornby DP (1996) Activation of a yeast pseudo DNA methyltransferase by deletion of a single amino acid. J Mol Biol 257:804–813

Posfai J, Bhagwat AS, Posfai G, Roberts RJ (1989) Predictive motifs derived from cytosine methyltransferases. Nucl Acids Res 17:2421–2435

Razin A, Cedar H (1994) DNA methylation and genomic imprinting. Cell 77:473–476

Razin A, Riggs AD (1980) DNA methylation and gene function. Science 210:604–610

Riggs AD, Pfeifer GP (1992) X-chromosome inactivation and cell memory. Trends Genet 8:169–174

Rougier N, Bourc'his D, Molina Gomes D, Niveleau A, Plachot M, Pàldi A, Viegas-Péquignot E (1998) Chromosome methylation patterns during mammalian pre-implantation development. Genes Dev 12.2108–2113

Sakai T, Toguchida J, Ohtani N, Yandell DW, Rapaport JM, Dryja TP (1991) Allele-specific hypermethylation of the retinoblastoma tumor-suppressor gene. Am J Hum Genet 48:880–888

Sanford JP, Clark HJ, Chapman VM, Rossant J (1987) Differences in DNA methylation during oogenesis and spermatogenesis and their persistence during early embryogenesis in the mouse. Genes and Devel 1:1039–1046

Shapiro LJ, Mohandas T (1982) DNA methylation and the control of gene expression on the human X chromosome. Cold Spring Harbor Symp Quant Biol 47:631–637

Sved J, Bird A (1990) The expected equilibrium of the CpG dinucleotide in vertebrate genomes under a mutation model. Proc Natl Acad Sci USA 87:4692–4696

Tate PH, Bird AP (1993) Effects of DNA methylation on DNA-binding proteins and gene expression. Curr Op Gen Dev 3:226–231

Tate P, Skarnes W, Bird A (1996) The methyl-CpG binding protein MeCP2 is essential for embryonic development in the mouse. Nature Genetics 12:205–208

Tucker KL, Beard C, Dausman J, Jackson-Grusby L, Laird PW, Lei H, Li E, Jaenisch R (1996a) Germ-line passage is required for establishment of methylation and expression patterns of imprinted but not of non-imprinted genes. Genes Dev 10:1008–1020

Tucker KL, Talbot D, Lee MA, Leonhardt H, Jaenisch R (1996b) Complementation of methylation deficiency in embryonic stem cells by a DNA methyltransferase minigene. Proc Natl Acad Sci USA 93:12920–12925

Turner BM (1991) Histone acetylation and control of gene expression. J Cell Sci 99:13–20

Wade PA, Jones PL, Vermaak D, Wolffe AP (1998) A multiple-subunit Mi-2 histone deacetylase from *Xenopus laevis* cofractionates with an associated Snf2 superfamily ATPase. Curr Biol 8:843–846

Wade PA, Gegonne A, Jones PL, Ballestar E, Aubry F, Wolffe AP (1999) Mi-2 complex couples DNA methylation to chromatin remodelling and histone deacetylation. Nature Genetics 23:62–66

Watt F, Molloy PL (1988) Cytosine methylation prevents binding to DNA of a HeLa cell transcription factor required for optimal expression of the adenovirus late promoter. Genes Dev 2:1136–1143

Wilkinson CRM, Bartlett R, Nurse P, Bird AP (1995) The fission yeast gene *pmt1* encodes a DNA methyltransferase homologue. Nucl Acids Res 23:203–210

Wyatt GR (1950) Occurrence of 5-methyl-cytosine in nucleic acids. Nature 166:237–238

Xue Y, Wong J, Moreno GT, Young, MK, Côté J, Wang W (1998) NURD, a novel complex with both ATP-dependent chromatin remodeling and histone deacetylase activities. Mol Cell 2:851–861

Yoder JA, Bestor TH (1998) A candidate mammalian DNA methyltransferase related to pmt1p of fission yeast. Hum Mol Genet 7:279–284

Yoder JA, Yen R-Y, Vertino PM, Bestor TH, Baylin SB (1996) New 5' regions of the murine and human genes for DNA-(cytosine-5)-methyltransferase. J Biol Chem 271:31092–31097

Zhang Y, LeRoy G, Seelig H-P, Lane WS, Reinberg D (1998) The dermatomyositis-specific autoantigen Mi2 is a component of a complex containing histone-deacetylase- and nucleosome-remodeling activites. Cell 95:279–289

Zhang Y, Ng HH, Erdjument-Bromage H, Tempst P, Bird A, Reinberg D (1999) Analysis of the NuRD subunits reveals a histone deacetylase core complex and a connection with DNA methylation. Genes Dev 13:1924–1935

Zuo S, Boorstein RJ, Teebor GW (1995) Oxidative damage to 5-methylcytosine in DNA. Nucl Acids Res 23:3239–3243

# Relationship Between Transcription and DNA Methylation

M.F. Chan, G. Liang, and P.A. Jones

## 1 Introduction

The CpG sequence is depleted about fivefold in the human genome. There are, however, regions of the genome that maintain a normal density of CpG dinucleotides. These CpG-rich regions are called CpG islands and have been defined as regions of DNA that are at least 200bp in length, have a G + C content greater than 0.5, and have an observed/expected presence of CpG greater than 0.6 (Gardiner-Garden and Frommer 1987). CpG islands are frequently found at the 5' ends of genes. Genes with wide-spread expression often have CpG islands starting 5' to the transcription unit and extending into one or more exons. In contrast, only about 60% of genes with a limited expression pattern have CpG islands that span the transcriptional start site. What is often overlooked is the fact that CpG islands can also occur in the coding regions of genes (Fig. 1), as found in the other 40% of genes with limited expression patterns (Larsen et al. 1992). Not much is known about the functional significance of these downstream islands. The focus of this

Department of Biochemistry and Molecular Biology, University of Southern California/Norris Comprehensive Cancer Center and Hospital, 1441 Eastlake Avenue, Room 83021, Mail Code 9181, Los Angeles, CA 90089-9181, USA
E-mail: jones_p@ccnt.hsc.usc.edu

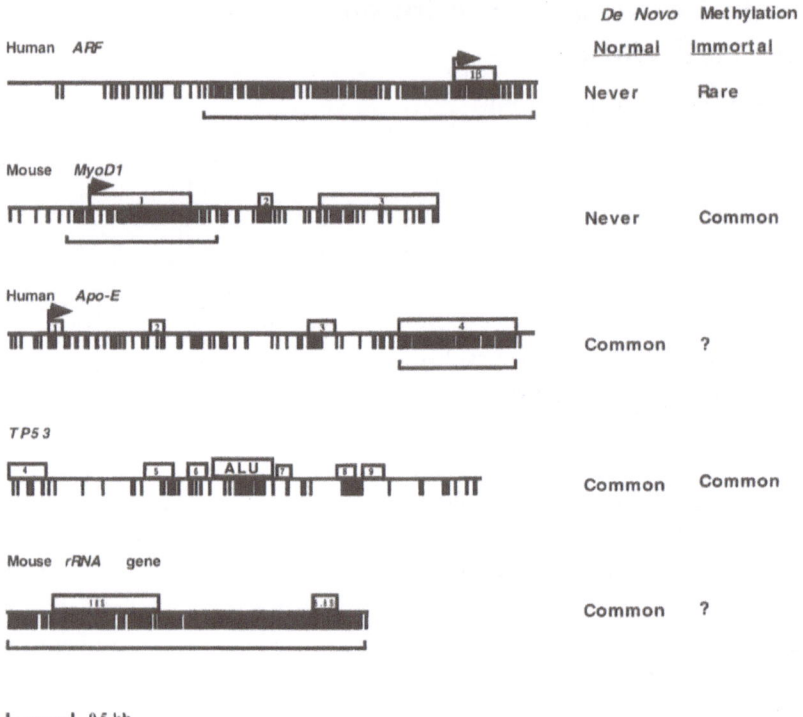

**Fig. 1.** Various Locations of CpG islands in Mammalian genes. CpG islands, bracketed in the above examples, can occur upstream of the transcriptional start site (*ARF*), around the region of the transcriptional start site (*MyoD1*), downstream of transcriptional start (*Apo-E*), or throughout a gene (*rRNA*). Regions of concentrated CpG sites, which do not always satisfy the criteria for CpG frequency to be called CpG islands, can occur within repetitive elements embedded in genes (*TP53*)

chapter will be to explore the relationship between CpG-island methylation and transcription. In particular, a comparison will be made between promoter-associated CpG islands and transcribed islands. Methylation at the site of transcription initiation can result in lack of expression, while methylation in downstream transcribed regions does not block the formation of a transcript.

## 2 Transcription of Genes with Downstream CpG Islands

Many lines of evidence support the idea that methylation of promoter-associated CpG islands correlates inversely with gene expression. One recently elucidated mechanism explaining how methylation might inhibit promoter activity will be discussed in detail in another chapter. Briefly, methylated-DNA-binding proteins binding methylated CpG islands can inhibit transcription by restricting the access of transcription factors and by recruiting a transcriptional repression complex involving histone deacetylases.

Because methylation of CpG-rich promoters has been shown to limit their activities, it has generally become accepted that methylation of CpG islands is associated with lack of transcription. However, this generalization should only be applied to promoter methylation. Unlike their promoter-associated counterparts, methylation of downstream CpG islands does not seem to inhibit the formation of a transcript. In fact, a close examination of existing methylation studies reveals that the degree of methylation of downstream CpG islands is often positively correlated with an increased level of gene expression (Table 1).

Housekeeping genes have vital functions and are expressed in virtually all cells. Therefore, it is not surprising that CpG islands are always associated with the 5′ ends of housekeeping genes (LARSEN et al. 1992). These 5′ CpG islands are unmethylated, which allows for gene expression. Methylation downstream of these

**Table 1.** Downstream methylation does not block transcription

| Gene | Location of transcribed CpG | Species | Timing | Reference |
|------|------------------------------|---------|--------|-----------|
| Methylation in transcribed regions of genes | | | | |
| aprt | Body of gene | Chinese hamster | Adult somatic cells | STEIN et al. 1983 |
| dhfr | Body of gene | Mouse | Adult somatic cells | STEIN et al. 1983 |
| α2-collagen | Exons 20–27 and exons 42–46 | Chick | Adult somatic cells | MCKEON et al. 1982 |
| apo-E | Exon 4 | Human | Adult somatic cells | LARSEN et al. 1993 |
| GFAP | Exon 1 | Rat | Embryonic brain cells | TETER et al. 1996 |
| rRNA | Body of gene | Mouse | Adult somatic cells | BROCK and BIRD 1997 |
| PAX6 | Exon 5 | Human | Cancer | LIANG et al. 1998 |
| p16 | Exon 2 | Human | Cancer (bladder) | GONZALGO et al. 1998 |
| PTHrP | Promoter 2 | Human | Cancer (lung) | GANDERTON and BRIGGS 1997 |
| p14 | Exon 2 | Human | Cancer (bladder) | GONZALGO et al. 1998 |
| Increased methylation with increased gene expression | | | | |
| hprt | Exon 3–9 | Mouse | Adult somatic cells | LOCK et al. 1986 |
| G6PD | Body of gene | Human | Embryonal carcinomas | BATTISTUZZI et al. 1985 |
| Igf2r | Intron 2 | Mouse | Blastocyst | WUTZ et al. 1997 |
| SNRPN | Intron 5 | Human | Peripheral blood | GLENN et al. 1996 |
| IGF2 | P2–P4 promoters | Human | Aging colonic epithelium | ISSA et al. 1996 |
| H-2K | Body of gene | Human | Embryonal carcinomas | TANAKA et al. 1983 |
| BCR-ABL | PA promoter | Human | Cancer (colon, bladder, prostate) | ZION et al. 1994 |
| BCL3 | Exons 4–7 | Human | Cancer (B-cell) | MCKEITHAN et al. 1994 |
| GPH | Exons 1 and 3 | Human | Cancer (various) | COX et al. 1998 |
| HOX | Body of gene | Human | Cancer (lung) | FLAGIELLO et al. 1996 |

ABL, Abelson leukemia; apo, apolipoprotein; aprt adenine phosphoribosyltransferase; BCL, B cell leukemia; BCR, breakpoint-cluster region; dhfr, dihydrofolate reductase; G6PD, glucose-6-phosphate dehydrogenase; GFAP, glial fibrillary acidic protein; GPH, glycoprotein hormone; HOX, homeobox-containing gene; hprt, hypoxanthine phosphoribosyltransferase; IGF2, insulin-like growth factor II; Igf2r, insulin-like-growth-factor-2 receptor; PTHrP, parathyroid hormone-related protein; rRNA, ribosomal RNA; SNRPN, small nuclear ribonucleoprotein polypeptide N.

islands does not seem to inhibit this expression, as is demonstrated by the Chinese-hamster adenine phosphoribosyltransferase (*aprt*) and mouse dihydrofolate reductase (*dhfr*) housekeeping genes. For both of these genes, the 5' region is unmethylated, while the 3' end is partially or fully methylated (STEIN et al. 1983).

This phenomenon of downstream methylation in genes with unmethylated 5' CpG islands is also observed in tissue-specific genes. The methylation status throughout the entire α2 (type-I) collagen gene was studied in cell types that synthesized either high, low, or no detectable levels of collagen. In all cell types, the region surrounding the start site of transcription was unmethylated, while the central and 3' regions of the gene were methylated. Therefore, α2 (type-I) collagen can still be transcribed even if segments of the gene are methylated (McKEON et al. 1982).

Not all genes have a CpG island associated with the transcriptional start site. Human apolipoprotein (*apo*)-E, with a CpG island located in exon 4, is an example of a gene with an internal CpG island (Fig. 1). This gene was found to be transcribed by most tissues even though its CpG island was fully methylated (LARSEN et al. 1993). Another internal CpG island is positioned upstream of the α-globin-cluster regulatory domain. While the associated gene was found to be constitutively expressed from a variety of cell lines and tissues, this 3' CpG island was found to be completely methylated in all samples studied, except for one partially methylated sample (VYAS et al. 1992). These are two more examples showing that downstream methylation is permissive for gene expression.

# 3 Methylation of Active Alleles During Development

It is clear that DNA methylation is essential for normal embryonic development. Mouse embryos homozygous for a DNA methyltransferase mutation die between 8.5–10.5 days postcoitum (LI et al. 1992). The pattern of methylation is very dynamic during embryonic development, with global demethylation occurring from the eight-cell stage to the blastocyst stage in mouse pre-implantation embryos, followed by a wave of *de novo* methylation at implantation.

## 3.1 X-Chromosome Inactivation

One of the earliest occurrences of *de novo* methylation of CpG islands occurs during pre-implantation development. At this time, X-chromosome inactivation occurs to ensure that female and male cells express equivalent levels of most X-linked gene products. This process is thought to involve three steps: an initial event in which one of the two X chromosomes is chosen for inactivation, the propagation of inactivation along the chromosome, and the heritable maintenance of the repression of genes on the inactive chromosome (GARTLER and RIGGS 1983). The process is initiated at the X-chromosome inactivation center (Xic; PETTIGREW et al. 1991).

A gene encoding the RNA molecule Xist maps to Xic and is transcribed monoal-lelically from the inactive X. It has been proposed that Xist plays a role in the initiation and spread of inactivation, because Xist expression is low prior to in-activation but becomes greatly elevated with an inactive X. It is well established that DNA methylation plays an important role in the propagation and mainte-nance of X inactivation (JAENISCH et al. 1998).

Early evidence showed that treatment of cells with the demethylating agent 5-azacytidine resulted in the reactivation of genes on the inactive X chromosome (MONHANDAS et al. 1981). In X-linked housekeeping genes, 5′ CpG islands are hypermethylated on the inactive X chromosome due to a wave of *de novo* met-hylation and are unmethylated on the active X chromosome. Therefore, the early transcription of Xist seems to facilitate the later *de novo* methylation of CpG islands on the same chromosome (MONK et al. 1995).

While much attention has been paid to the fact that 5′ CpG-island methylation in the X-linked genes is often correlated with gene inactivity, less focus has been placed on the finding that several of these genes, including mouse hypoxanthine phosphoribosyltransferase (*hprt*) and human glucose-6-phosphate dehydrogenase (*G6PD*), also contain downstream methylated regions that are positively associated with gene expression. The mouse *hprt* gene contains two regions of differential methylation. One region involves the 5′ region of the gene that contains CpG sites that are completely unmethylated when carried on the active X and that are ex-tensively methylated when carried on the inactive X. There is also a second region of differential methylation that is present at the 3′ end of the gene. Contrary to what was found at the 5′ end of the gene, sites in the 3′ region are completely methylated when carried on the active X and are always unmethylated when carried on the active X (LOCK et al. 1986). The human *G6PD* gene has two regions with differing methylation patterns similar to that of the mouse *hprt* gene. The methylation of specific CpG sites in the 5′ region are again methylated on the inactive X chro-mosome. A study of the extent of 3′ CpG methylation between different tissues with varying *G6PD* expression revealed that tissues with the highest levels of expression also had the highest levels of methylation (BATTISTUZZI et al. 1985).

Similar to the mouse *hprt* and human *G6PD* genes, the hypervariable DXS255 locus contains a sequence that is extensively methylated on the active X and es-sentially unmethylated on the inactive X. Interestingly, this sequence is part of a CpG island at the 5′ end of a long interspersed repetitive element (LINE-1) (HENDRIKS et al. 1992). This observation supports the possibility that LINE-1 el-ements are only potentially harmful in transcriptionally active regions of the ge-nome and that transcription helps lead to their methylation.

## 3.2 Imprinting

One of the best examples of a causal link between transcription and methylation is the mouse imprinted gene insulin-like-growth-factor-2-receptor (*Igf2r*) gene. This gene contains, in its second intron, a CpG island referred to as region 2, which is thought to be an imprinting signal. Transcription of the maternal allele results in the *de novo* methylation of region 2 and consequently inhibits its ability to act as a promoter for antisense RNA. Transcription of the paternal allele produces the opposite pattern – region 2 is unmethylated, and an antisense messenger (mRNA) transcript is formed. Interestingly, active transcription of this antisense RNA runs through the upstream CpG island, and methylation of the *Igf2r* promoter results (along with the absence of *Igf2r* mRNA). To further support the role of transcription in *de novo* methylation, transgenes that were inadvertently derived with inactive promoters failed to *de novo* methylate region 2 after maternal or paternal inheritance (WUTZ et al. 1997).

The imprinted gene, human small nuclear ribonucleoprotein polypeptide N (*SNRPN*), has a methylation pattern very similar to that of *Igf2r*. However, in this case, the transcriptional start site that lies within a CpG island is methylated in the repressed, maternally inherited allele while the expressed, paternally inherited promoter remains unmethylated. *De novo* methylation occurs about 22kb downstream of the start site, within intron 5 of the expressed allele, and remains unmethylated in the silent, maternally inherited allele (GLENN et al. 1996).

Another imprinted gene, human insulin-like growth factor II (*IGF2*), has a similar link between transcription and methylation in the colonic epithelium of aging individuals. It is transcribed from four promoters, P1–P4. The P2–P4 promoters lie within a CpG island and are paternally expressed. The P1 promoter lies upstream of the CpG island and is regulated in a different manner. A study of the methylation state of the P2–P4 CpG island in normal tissues revealed that methylation of the island was particularly extensive in the liver. The liver is also the tissue that has the highest level of *IGF2* expression driven by the P1 promoter. The possible link between expression of the P1 promoter and methylation of the CpG island was also observed in cell lines. Cell lines that transcribed the upstream P1 promoter were prone to hypermethylate *IGF2* (ISSA et al. 1996).

These three imprinting examples provide clear evidence that *de novo* methylation of CpG islands downstream of the transcriptional start site is correlated with expression. In all three cases, the downstream sequences became *de novo* methylated in the transcribed allele, while the silent allele remained unmethylated. Therefore, transcription through a sequence may facilitate its methylation.

## 3.3 Tissue-Specific Genes

DNA in normal, differentiated somatic cells has a fixed pattern of methylation that is faithfully copied after replication. In contrast, the methylation patterns of DNA from several tissue-specific and housekeeping genes have been found to change during normal development.

An important set of genes that becomes expressed just after the implantation stage of development are the class-I major histocompatibility complex (MHC) genes, which encode the classical transplantation antigens. The increased methylation of the H2K promoter in one of the two homologous chromosomes was accompanied by a low level of expression of the H2K antigen, while an increased methylation in both chromosomes was accompanied by a high level of expression. Therefore, the activation of the class-I MHC antigen H2K was found to be positively correlated with an increase in methylation (TANAKA et al. 1983).

The methylation pattern of the glial fibrillary acidic protein (*GFAP*) gene in the rat has been found to change during brain development. However, this change is not uniform throughout the gene. While the promoter showed a wave of demethylation during the onset of *GFAP* transcription in the embryo, followed by re-methylation at later postnatal stages, the downstream CpG island remained highly methylated throughout the development of the rat's brain (TETER et al. 1996). This is clear evidence that methylation of the promoter and coding sequence can be differentially regulated during development and that methylation of the coding sequence can still allow for the expression of developmentally regulated genes.

# 4 Transcription Through Methylated Repetitive Elements

The genome-defense model suggests that DNA methylation serves a similar role in protecting animal genomes (BESTOR 1998). At least 35% of the human genome is composed of transposable elements and repetitive sequences, such as LINEs and short interspersed repetitive elements (YODER et al. 1997). CpG-rich Alu sequences are often found less than 1 kb from CpG islands or at the borders of islands (LARSEN et al. 1992). It is important that these elements not be transcriptionally active, because of their potential detrimental effects on the genome. Transposable elements threaten the structure and the expression of the genome through insertion mutations, translocations and chimeric mRNAs originating at the promoters of retrotransposons (YODER et al. 1997). These sequences have been found to be targets of *de novo* methylation. Infection of embryonic stem cells homozygous for a null mutation of the maintenance methyltransferase *DnmtI* with the murine retrovirus MoMuLV can result in the *de novo* methylation of the integrated provirus (LEI et al. 1996).

Alu elements can be located in close association with growth-regulating genes, such as *p53* and *p16*. The Alu elements in intron 6 of *p53* and in the region between exons 1δ and 1β of *p16* are heavily methylated in all cell types examined. Our laboratory has demonstrated that transcription of *p53*, *p16* and *p14* can still occur despite the heavy methylation of the intervening repetitive elements (MAGEWU et al. 1994; GONZALGO et al. 1998)

Ribosomal RNA (rRNA) genes are present in multiple copies in the genomes of many species. In the human sequence, the 43-kb repeat unit is divided into a 13.3-kb transcribed region which encodes the 18S, 5.8S, and 28S rRNA subunits and a 30-kb non-transcribed spacer. CpG islands span their entire transcriptional unit. In general, the transcribed unit is unmethylated, while the non-transcribed spacer is methylated in humans. However, this is not true across all species, as it has been demonstrated that rRNA is heavily methylated in both transcribed and non-transcribed regions of the repeat unit in amphibians and fish. This phenomenon is also observed in mice, as a small proportion of all repeat units are methylated throughout the transcribed region (BROCK and BIRD 1997). Therefore, in these species, methylation does not block RNA-polymerase-I activity.

# 5 Aberrant Methylation Patterns in Cancer

Over the past few years, there has been increasing support for the idea that methylation plays a role in cancer. Alterations in global levels and regional changes in the patterns of DNA methylation are among the earliest and most frequent events known to occur in human cancers (JONES 1986). At the regional level, these changes may impair the correct expression or function of cell-cycle-regulatory genes and, consequently, give affected cells a selective growth advantage. A more global reduction of methylation may allow for a decrease in chromosomal stability (CHEN et al. 1998). Abnormal hypermethylation of CpG islands in the promoters of several tumor-suppressor genes, including retinoblastoma (SAKAI et al. 1991), Von-Hippel Lindau (HERMAN et al. 1994) and Wilms' Tumor genes (STEENMAN et al. 1994), leads to their inactivation and has been found in several tumors.

Methylation does not always abrogate expression in tumors. In fact, breakpoint-cluster region–Abelson leukemia (*BCR-ABL*) exhibits the phenomenon of de novo methylation with active transcription. In human chronic myelogenous leukemia, a chromosomal translocation event causes most of the proto-oncogene *ABL* to be fused to the 5' region of the *BCR* gene. This hybrid gene is regulated by the *bcr* promoter, with the *ABL* promoter (Pa) being nested within a CpG island in the transcriptional unit. During the course of the disease, the Pa promoter undergoes progressive *de novo* methylation on the Ph' chromosome, with increasing *BCR-ABL* expression, while the normal *ABL* allele remains hypomethylated (ZION et al. 1994). In tumors from patients with colon or bladder or prostate cancer, hypermethylation of the CpG island in exon 5 of the *PAX6* gene is associated with increased expression (Saiem et al.). The same positive correlation between expression and *de novo* methylation was observed in *p16* exon 2 of bladder and colon tumors (GONZALEZ-ZULUETA et al. 1995).

Abnormal hypermethylation in tumors is not restricted to promoters and has also been found elsewhere in genes. The candidate proto-oncogene, B cell leukemia 3 (*BCL3*), is involved in a translocation event found in some cases of human B cell chronic lymphocytic leukemia or other B-cell neoplasms. This gene contains both a 5' CpG island that encompasses the first exon and another, more 3' CpG island that spans about 1.6kb of sequence, including exons 4–7. In all tissues studied, the 5' CpG island was unmethylated. This contrasts with the 3' CpG island, which was completely unmethylated in sperm DNA, completely methylated in DNA from peripheral blood leukocytes of a patient with chronic myelogenous leukemia and partially methylated in DNA from other tissues (McKEITHAN et al. 1994).

The glycoprotein hormone α-subunit (GPHα) is another gene with two regions of CpG-rich sequences. This gene contains Alu repetitive sequences in the 5'-flanking DNA and in the second intron. Interestingly, both CpG-rich sequences were hypermethylated in a variety of tumor cell lines that produced α-subunit at high levels but were significantly less methylated in tumor cell lines where the gene was unexpressed or expressed at low levels. In this example, methylation seems to increase with increased expression (Cox et al. 1998).

The parathyroid hormone-related protein (*PTHrP*) gene is under the control of three distinct promoters. A CpG island lies between promoters 1 and 2. Two-thirds of this CpG island was substantially methylated in the lung squamous cell carcinoma cell line, BEN. Despite this heavy methylation, expression was not inhibited from any of the three gene promoters (GANDERTON and BRIGGS 1997). This is an interesting example of how methylation of a CpG island does not inhibit transcription even though its relative position to the three promoters differs.

*p16* is a cell-cycle-regulatory gene which has CpG islands in its promoter and second exon. The *p14* cell-cycle-regulatory gene also has a CpG island in its promoter, and this promoter originates about 20kb upstream of the *p16* exon-1α-coding domain. The *p16* and *p14* transcripts use alternative first exons ($1\alpha + 1\beta$) joined through the same splice acceptor site to exon 2 but in different reading frames. Our lab has found that, in bladder-cancer cell lines where the 5' CpG island of the *p16* gene is methylated, the *p14* transcriptional machinery is able to pass through this highly CpG-rich sequence and generate a *p14* transcript. Therefore, while methylation of the *p16* 5' CpG island blocks the expression of the *p16* transcript, it does not block the formation of the *p14* transcript (GONZALGO et al. 1998).

The human homeobox-containing (*HOX*) genes are part of a family of evolutionarily conserved transcription factors. Available sequences indicate that the *HOX* loci are CpG-rich regions. The 38 members of the family are tandemly arranged in four clusters on chromosomes 7 (*HOXA*), 17 (*HOXB*), 12 (*HOXC*) and 2 (*HOXD*). The *HOXB* locus contains a row of nine *HOX* genes, and its expression was found to vary between different xenografted small-cell lung cancers. Four of the nine genes followed the rule "methylation expression, no methylation no expression" in an analysis of the 3' end of the genes. The other five genes also followed this rule in a few tumor samples, but exceptions were found in other tumor samples (FLAGIELLO et al. 1996).

# 6 Does Transcription Facilitate De Novo Methylation?

It is quite clear that 3' CpG islands seem to be regulated differently from 5' CpG islands, because the methylation status of each often has an opposite correlation with transcription. MACLEOD et al. (1998) have suggested that all CpG islands exist as promoters that are active in early embryonic development. They propose that the binding of transcription factors to CpG-island promoter sequences protects them from the wave of global *de novo* methylation that occurs during gametogenesis and mammalian embryonic development. This suggestion is supported by the fact that Sp1 elements are able to protect CpG islands from *de novo* methylation (BRANDEIS et al. 1994).

84     M.F. Chan et al.

Based on this embryonic origin of CpG islands and the fact that transcribed CpG islands are often methylated, we propose the model shown in Fig. 2. In this model, protein factors that are bound to CpG islands (such as Sp1) protect CpG islands from the wave of *de novo* methylation occurring during early development. Following this event, transcription initiating upstream of a CpG island results in the movement of a transcriptional complex through it. The DNA takes on an open conformation, making it a target for methylation by a *de novo* methyltransferase. Maintenance methylation can methylate the other DNA strand, and transcription factors are now less able to bind this methylated form of DNA. This initial seeding of methylation can then lead to methylation spreading throughout the CpG island.

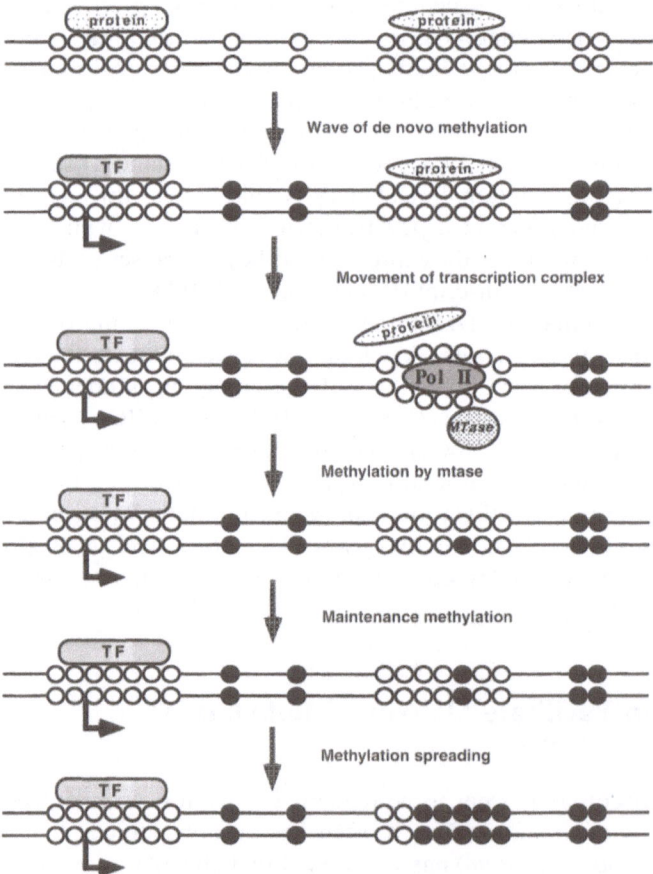

**Fig. 2.** Model for the methylation of downstream CpG islands. A gene containing two CpG islands is shown. Both islands are protected from the wave of de novo methylation during development because they are bound to protein factors. The downstream CpG island becomes de novo methylated as a result of a transcriptional complex opening up the DNA, because the open DNA form attracts methyltransferase activity. Maintenance methylation methylates the other strand, and this methylation becomes the nidus for the spreading of methylation. Unmethylated CpG sites are depicted as *open circles*, and methylated sites are shown as *filled circles*. Transcription start sites are shown as *arrows*

According to this model, *de novo* methylation may be more pronounced in cancer cells because of the transcription of genes at an increased rate or in an inappropriate cell context. This inappropriate methylation of genes can lead to a growth advantage for the tumor cell only if it occurs in the promoter of a growth-regulatory gene.

# 7 Conclusions

The effects of CpG-island methylation on transcription depend on the location of the CpG island in the transcriptional unit. While the mechanism by which promoter methylation leads to gene inactivity is becoming clearer, the significance of downstream CpG-island methylation remains unclear. The many examples of transcribed CpG-island methylation in development, somatic cells, and cancer support the fact that downstream methylation is the rule rather than the exception. Therefore, it is only when we have a better understanding of downstream methylation that we can begin to identify some of the underlying mechanisms of *de novo* methylation.

# References

Battistuzzi G, D'Urso M, Toniolo D, Persico GM, Luzzatto L (1985) Tissue-specific levels of human glucose-6-phosphate dehydrogenase correlate with methylation of specific sites at the 3' end of the gene. Proc Natl Acad Sci 82:1465–1469

Bestor TH (1998) In: Chadwick DJ, Cardew G (eds) The host defence function of genomic methylation patterns. Epigenetics. Novartis Foundation, West Sussex, pp 187–194

Brandeis M, Frank D, Keshet I, Siegfried Z, Mendelsohn M, Nemes A, Temper V, Razin A, Cedar H (1994) Sp1 elements protect a CpG island from *de novo* methylation. Nature 371:435–438

Brock GJ, Bird A (1997) Mosaic methylation of the repeat unit of the human ribosomal RNA genes. Hum Mol Genet 6:451–456

Chen RZ, Pettersson U, Beard C, Jackson-Grusby L, Jaenisch R (1998) DNA hypomethylation leads to elevated mutation rates. Nature 395:89–93

Cox GS, Gutkin DW, Haas MJ, Cosgrove DE (1998) Isolation of an Alu repetitive DNA binding protein and effect of CpG methylation on binding to its recognition sequence. Biochim Biophys Acta 1396:67–87

Flagiello D, Poupon M, Cillo C, Dutrillaux B, Malfoy B (1996) Relationship between DNA methylation and gene expression of the HOXB gene cluster in small cell lung cancers. FEBS 380:103–107

Ganderton RH, Briggs RSJ (1997) CpG-island methylation and promoter usage in the parathyroid hormone-related protein gene of cultured lung cells. Biochim Biophys Acta 1352:303–310

Gardiner-Garden M, Frommer M (1987) CpG islands in vertebrate genomes. J Mol Biol 196:261–282

Gartler SM, Riggs AD (1983) Mammalian X-chromosome inactivation. Annu Rev Genet 17:155–190

Glenn CC, Saitoh S, Jong MTC, Filbrandt MM, Surti U, Driscoll DJ, Nicholls RD (1996) Gene structure, DNA methylation, and imprinted expression of the human SNRPN gene. Am J Hum Genet 58:335–346

Gonzalez-Zulueta M, Bender CM, Yang AS, Nguyen T, Beart RW, Van Tornout JM, Jones PA (1995) Methylation of the 5' CpG island of the p16/CDKN2 tumor suppressor gene in normal and transformed human tissues correlates with gene silencing. Cancer Res 55:4531–4535

Gonzalgo ML, Hayashida T, Bender CM, Pao MM, Tsai YC, Gonzales FA, Nguyen HD, Nguyen TT, Jones PA (1998) The role of DNA methylation in the expression of the *p19/p16* locus in human bladder cancer cell lines. Cancer Res 58:1245–1252

Hendriks RW, Hinds H, Chen Z, Craig IW (1992) The hypervariable DXS255 locus contains a LINE-1 repetitive element with a CpG island that is extensively methylated only on the active X chromosome. Genomics 14:598–603

Herman JG, Latif F, Weng Y, Lerman MI, Zbar B, Liu S, Samid D, Duan DS, Gnarra JR, Linehan WM, et al. (1994) Silencing of the VHL tumor-suppressor gene by DNA methylation in renal carcinoma. Proc Natl Acad Sci 91:9700–9704

Issa JJ, Vertino PM, Boehm CD, Newsham IF, Baylin SB (1996) Switch from monoallelic to biallelic human IGF2 promoter methylation during aging and carcinogenesis. Proc Natl Acad Sci 93:11757–11762

Jaenisch R, Beard C, Lee J, Marahrens Y, Panning B (1998) Mammalian X-chromosome inactivation. In: Chadwick DJ, Cardew G (eds) Epigenetics. Novartis Foundation, West Sussex, pp 200–208

Jones PA (1986) DNA methylation and cancer. Cancer Res 46:461–466

Larsen F, Gundersen G, Lopez R, Prydz H (1992) CpG islands as gene markers in the human genome. Genomics 13:1095–1107

Larsen F, Solheim J, Prydz H (1993) A methylated CpG island 3′ in the apolipoprotein-E gene does not repress its transcription. Hum Mol Genet 2:775–780

Lei H, Oh SP, Okano M, Jüttermann R, Goss KA, Jaenisch R, Li E (1996) De novo DNA cytosine methyltransferase activities in mouse embryonic stem cells. Development 122:3195–3205

Li E, Bestor TH, Jaenisch R (1992) Targeted mutation of the DNA methyltransferase gene results in embryonic lethality. Cell 69:915–926

Liang G, Salem CE, Yu MC, Nguyen HD, Gonzales FA, Nguyen TT, Nichols PW, Jones PA (1998) DNA-methylation differences associated with tumor tissues identified by genome scanning analysis. Genomics 53:260–8

Lock LF, Melton DW, Caskey CT, Martin GR (1986) Methylation of the mouse hprt gene differs on the active and inactive X chromosomes. Mol Cell Biol 6:914–924

Macleod D, Ali RR, Bird A (1998) An alternative promoter in the mouse major histocompatibility complex class-II I-Aβ gene: implications for the origin of CpG islands. Mol Cell Biol 18:4433–4443

Magewu AN, Jones PA (1994) Ubiquitous and tenacious methylation of the CpG site in codon 248 of the p53 gene may explain its frequent appearance as a mutational hot spot in human cancer. Mol Cell Biol 14:4225–4232

McKeithan TW, Ohno H, Dickstein J, Hume E (1994) Genomic structure of the candidate proto-oncogene BCL3. Genomics 24:120–126

McKeon C, Ohkubo H, Pastan I, de Crombrugghe B (1982) Unusual methylation pattern of the α2(I) collagen gene. Cell 29:203–210

Mohandas T, Sparkes RS, Shapiro LJ (1981) Reactivation of an inactive human X chromosome: evidence for X-inactivation by DNA methylation. Science 211:393–396

Monk M (1995) Epigenetic programming of differential gene expression in development and evolution. Dev Genet 17:188–197

Pettigrew A, Ledbetter AL, Levy DH, Craig IW, Willard HF (1991) Localization of the X inactivation center on the human X chromosome in Xq13. Nature 349:82–84

Saiem CE, Marki IDC, Bender CM, Gonzaies FA, Jones PA, Liang G (1999 submitted) PAX6 Methylation and ectopic expression in human tumor cells

Sakai T, Toguchida J, Ohtani N, Yandell DW, Rapaport JM, Dryja TP (1991) Allele-specific hypermethylation of the retinoblastoma tumor-suppressor gene. Am J Hum Genet 48:880–888

Steenman MJ, Rainier S, Dobry CJ, Grundy P, Horon IL, Feinberg AP (1994) Loss of imprinting of IGF2 is linked to reduced expression and abnormal methylation of H19 in Wilms' tumour. Nat Genet 7:433–439

Stein R, Sciaky-Gallili N, Razin A, Cedar (1983) Pattern of methylation of two genes coding for housekeeping functions. Proc Natl Acad Sci 80:2422–2426

Tanaka K, Appella E, Jay G (1983) Developmental activation of the H-2 K gene is correlated with an increase in DNA methylation. Cell 35:457–465

Teter B, Rozovsky I, Krohn K, Anderson C, Osterburg H, Finch C (1996) Methylation of the glial fibrillary acidic protein gene shows novel biphasic changes during brain development. Glia 17:195–205

Vyas P, Vickers MA, Simmons DL, Ayyub H, Craddock CF, Higgs DR (1992) Cis-acting sequences regulating expression of the human α-globin cluster lie within constitutively open chromatin. Cell 69:781–793

Wutz A, Smrzka OW, Schweifer N, Schellander K, Wagner EF, Barlow DP (1997) Imprinted expression of the Igf2r gene depends on an intronic CpG island. Nature 389:745–749

Yoder JA, Walsh CP, Bestor TH (1997) Cytosine methylation and the ecology of intragenomic parasites. Trends Genet 13:335–340

Zion M, Ben-Yehuda D, Avraham A, Cohen O, Wetzler M, Melloul D, Ben-Neriah Y (1994) Progressive de novo DNA methylation at the bcr-abl locus in the course of chronic myelogenous leukemia. Proc Natl Acad Sci 91:10722–10726

# DNA Methylation, Genomic Imprinting and Cancer

A.P. FEINBERG

## 1 Introduction

It has been 16 years since the discovery of altered gene methylation in human cancer (FEINBERG and VOGELSTEIN 1983) and 6 years since abnormal imprinting in cancer was discovered (RAINIER et al. 1993), but the relationship between these two types of epigenetic alteration in molecular oncology is only beginning to be understood. The topic of genomic imprinting and, in particular, the role of genomic imprinting in cancer, have been amply reviewed elsewhere, including two recent reviews by this author (FEINBERG 1999). Therefore, as this is a volume on DNA methylation, I will focus this chapter on the relationship between DNA methylation and altered imprinting in cancer. However, some discussion of imprinting in general is necessary for clarity.

Genomic imprinting is defined as a parental-origin-specific epigenetic modification of a gene or the chromosome on which it resides in the gamete or zygote, leading to differential expression of the two alleles in somatic cells of the offspring. There are several important properties embedded in this definition. First, parental-origin specificity is necessary to distinguish imprinting from allelic exclusion, as is seen in the immunoglobulin- and olfactory-receptor genes. Second, the definition does not require monoallelic expression, although it is often incorrectly assumed.

Johns Hopkins University School of Medicine, 720 Rutland Avenue, Baltimore, MD 21205, USA
E-mail: afeinberg@jhu.edu

Thus, the two alleles may both be expressed but expressed unequally. Third, a modification or mark must take place in the gamete or zygote, since that is the latest stage at which the two parental contributions to the genome are still topologically distinct. However, allele-specific expression itself might not arise until later in development. Alternatively, allele-specific expression could be present but then lost later in development. This is an important point that is also often overlooked, but it is directly relevant to a potential role for DNA methylation in genomic imprinting, since methylation also shows developmentally specific changes. Finally, the definition of imprinting given above has migrated somewhat from the original definition advanced by Helen Crouse. Studying the insect *Sciara*, Crouse observed preferential exclusion of paternal chromosomes in development, and her definition, therefore, referred to an epigenetic mark and was not related to gene expression at all (CROUSE 1960). Nevertheless, since the imprinting mark is (still) unknown but allele-specific gene expression has become easy to measure, the expression-based definition above is now generally accepted.

Genomic imprinting is important, because it indicates that the maternal and paternal genomes are not equivalent. For example, trisomic mouse and human embryos have markedly different phenotypes depending on the parental origin of the excess genome. Similarly, both mouse and human chromosomes that undergo uniparental disomy (duplication of one parental copy and loss of the other parental copy) often show characteristic phenotypic alterations in the offspring. These can include overgrowth in the case of paternal uniparental disomy for some chromosomal regions, and growth retardation in the case of maternal uniparental disomy for the same chromosomal regions (CATTANACH and BEECHEY 1990; LEDBETTER and ENGEL 1995). This parental-origin-specific difference led Haig to hypothesize that imprinting arose in evolution because of competition for nutrients between the mother and developing offspring in utero. According to this hypothesis, in polygamous mammals, the paternal genome tends to promote growth at the expense of other littermates and the future reproductive health of the mother, and the maternal genome tends to oppose this effect (HAIG and GRAHAM 1991; MOORE and HAIG 1991). To date, over 30 imprinted genes have been identified, including genes for growth, neurological development, growth factors and their receptors, ion channels, and genes involved in RNA processing (MORISON and REEVE 1998).

# 2 Genomic Imprinting and DNA Methylation

Curiously, recognition of a role for DNA methylation in imprinting began with the initial discovery of mammalian imprinting involving an artificially introduced transgene. SWAIN et al. (1987), REIK et al. (1987), and SAPIENZA ct al. (1987) all observed parental-origin-specific methylation of transgenes. For example, the paternally inherited *c-myc* transgene is relatively undermethylated and expressed in the heart, while the maternally inherited transgene is methylated and not expressed

(SWAIN et al. 1987). Transgene methylation was found to be erased and then re-programmed in primordial germ cells (SWAIN et al. 1987).

DECHIARA et al. (1991) were the first to discover an endogenous mammalian imprinted gene. This gene, insulin-like growth factor II (*Igf2*), has become a focus of many laboratories studying human cancer. Allele-specific expression of *Igf2* is partly but not entirely explained by regulation by the transcriptional unit of *H19*, an untranslated RNA that is expressed from the maternal allele (BARTOLOMEI et al. 1991). Targeted disruption of *Igf2* leads to dwarfism when transmitted from the father (DECHIARA et al. 1991), and targeted disruption of *H19* leads to overgrowth when transmitted from the mother (LEIGHTON et al. 1995). The first link between DNA methylation and imprinting of an endogenous gene was made by two groups who observed that a CpG island within and upstream of the *H19* gene is methylated on the paternal allele and unmethylated on the maternal allele (BARTOLOMEI et al. 1993; FERGUSON-SMITH et al. 1993). Methylation of the gene itself and its promoter extends to all tissues except for the sperm, in which it is hypomethylated (BART-OLOMEI et al. 1993; FERGUSON-SMITH et al. 1993). However, the approximately 4-kb region 5' to the promoter shows a similar pattern of allele-specific methylation that also extends to the sperm (BARTOLOMEI et al. 1993; TREMBLAY et al. 1995). Thus, this element could represent an initial methylation-specific imprinting mark, since it is present in the developing gamete. Evidence for such an idea was recently adduced from a targeted disruption of this element, which led to a reversal of the normal imprinting pattern in the offspring (THORVALDSEN et al. 1998). However, the fact that the imprinting pattern was reversed and not simply erased suggests that this element is necessary for propagation of an initial mark or is part of that initial mark (but not all of it).

Another gene for which a methylation-linked imprinting mark has been identified is the Igf2-receptor gene (*Igf2r*). Within the first intron of this gene lies a CpG island that is methylated on the maternal, *expressed* allele (STOGER et al. 1993). Thus, imprint-specific methylation is not necessarily associated with silencing. Targeted disruption of this element in a yeast artificial chromosome leads to loss of imprinting of *Igf2r* (WUTZ et al. 1997). Other methylated CpG islands showing parental-origin-specific methylation have been observed in the *Igf2* gene itself, including a region that is methylated on the *active* allele (SASAKI et al. 1992), *p57^{KIP2}* in mouse (HATADA et al. 1996), $K_VLQT1$ (which will be discussed later in this chapter), *Grf1* (PLASS et al. 1996), *Peg1/Mest* (RIESEWIJK et al. 1997), mouse *U2af1-rs1* (SHIBATA et al. 1996), and the *Snrpn* gene (BUITING et al. 1994; REIS et al. 1994). Deletion of a methylated CpG island in the human *SNRPN* gene leads to a failure of reprogramming of the *SNRPN* imprint in patients with Prader-Willi Syndrome, a disorder of movement, appetite, and intellectual function (BUITING et al. 1995). Altered imprinting extends for several megabases in both directions, including several other CpG islands that show altered DNA methylation, indicating that the consequences of this deletion have an effect on chromatin over a large distance (BUITING et al. 1995). However, it is not clear whether it is a failure of methylation or loss of expression of the *SNRPN* gene itself that is responsible for this effect.

Persuasive evidence for the importance of the intronic *Igf2r* CpG island in regulating imprinting comes from an artificial system in which constructs containing this region were introduced into the pronuclei of embryonic stem cells, leading to parental-origin-specific methylation (BIRGER et al. 1999). However, neither recapitulation of an imprinting pattern in vivo using this sequence in an exogenously introduced transgene nor deletion of the endogenous element has been performed as of this writing. As noted earlier, *YAC* transgenes containing the *Igf2r* region and, in other experiments, the *IGF2-H19* region, have been introduced (AINSCOUGH et al. 1997; WUTZ et al. 1997). These were both able to undergo normal imprinting, at least sometimes (AINSCOUGH et al. 1997; WUTZ et al. 1997). It seems that a large genomic region is necessary for something resembling physiological imprinting, since smaller transgenes have no (or inconsistent) effects or simply show maternal silencing, as was observed for the simple transgenes studied a decade ago (REIK et al. 1987; SAPIENZA et al. 1987; SWAIN et al. 1987). At the very least, specific CpG islands appear to be necessary for imprinting establishment and or maintenance, but they do not represent the whole story.

In addition to these experiments showing a role for methylation in *cis*-acting sequences, DNA-methyltransferase-deficient mice convincingly show a role for DNA methylation in *trans*, in this case caused by the methylation-maintenance machinery itself. As might be expected, such mice show loss of methylation of the CpG islands that normally exhibit allele-specific methylation associated with known imprinted genes (LI et al. 1993). Correspondingly, these same genes do not show normal imprinting (LI et al. 1993). Thus, DNA methylation does appear to be necessary for normal imprinting. However, that is not necessarily always the case, as the *Mash2* gene has been found to be normally imprinted in these mice (CASPARY et al. 1998). Thus, it seems likely that methylation-independent mechanisms are also important in establishing and maintaining normal genomic imprinting.

# 3 Genomic Imprinting and Cancer

A role for genomic imprinting in cancer has long been suspected, because whole-genome uniparental disomy leads to tumors. Thus, hydatidiform moles arises from androgenetic embryos with a normal chromosome number (KAJII and OHAMA 1977). Similarly, complete ovarian teratoma, a benign tumor, arises from gynogenetic embryos of normal chromosome number (LINDER et al. 1975). The complete ovarian teratoma is a benign tumor that can contain hair, teeth, and other tissues, while the hydatidiform mole is a relatively undifferentiated tumor with malignant potential. Clearly, an excess of maternal or paternal genome leads to neoplastic growth, and the type of growth depends on the parental chromosome in excess. Specific chromosomes that might be responsible for imprinting-related neoplasia were inferred from observations of parent-of-origin-specific chromosomal alterations in tumors. These include maternal- or paternal-specific loss of heterozygosity (LOH) and parental-origin-specific gene amplification (FEINBERG 1993). Because of parental-origin-specific LOH, it was hypothesized by Sapienza that the role of imprinting in cancer would be that some tumor-suppressor genes are imprinted and that the expressed allele is the allele undergoing LOH (SCRABLE et al. 1989; SAPIENZA 1990). This model required that the imprinting arise abnormally and mosaically during development, since otherwise an imprinted tumor-suppressor gene would behave as a Knudsonian two-hit recessive locus, in which one allele was always inactivated by virtue of its epigenetic silencing due to imprinting (SCRABLE et al. 1989; SAPIENZA 1990).

However, the first gene abnormality in cancer related to imprinting involved abnormal *activation* of the normally silent copy of two imprinted genes, *IGF2* (OGAWA et al. 1993; RAINIER et al. 1993) and *H19* (RAINIER et al. 1993). This was a completely unanticipated mechanism but, subsequently, both LOH and epigenetic silencing of the expressed allele of growth-inhibitory genes have also been described (STEENMAN et al. 1994; MATSUOKA et al. 1996; THOMPSON et al. 1996). The term that we coined for this novel epigenetic alteration in cancer was "loss of imprinting" or LOI, which simply means loss of the normal parental-origin-specific pattern of differential allele expression of an imprinted gene (FEINBERG 1993; RAINIER et al. 1993). Thus, LOI could involve activation of the normally silent allele (as seen in *IGF2*) or silencing of the normally expressed allele. LOI also need not require absolute erasure of an imprinting mark. It should be remembered that normally imprinted genes do not necessarily show allele-specific expression in all tissues at all stages of development. For example, *H19*, one of the archetypical imprinted genes, may not exhibit allele-specific expression in the early embryo (SZABO and MANN 1995) even though the gene is marked in the germline (TREMBLAY et al. 1995).

To date, all of the genes that have shown abnormal imprinting in cancer are located within a 1-Mb region of human chromosome 11p15 (HU et al. 1997; FEINBERG 1999). The investigation of this chromosomal region has provided insight into abnormal imprinting in cancer and into the organization and normal regulation of imprinted genes. Most initial efforts were focused on *IGF2*, which has now been shown to undergo altered imprinting in most human cancers (FEINBERG 1999). While the initial investigations of *IGF2* imprinting were performed on embryonal tumors of children, including Wilms' tumor (LI 1990; OGAWA et al. 1993), hepatoblastoma (RAINIER et al. 1995), and rhabdomyosarcoma (ZHAN et al. 1994), subsequent studies have also shown loss of imprinting in most common adult tumors, including hepatocellular (TAKEDA et al. 1996), cervical (ALVAREZ et al. 1995), colorectal (CUI et al. 1998), brain (UYENO et al. 1996) and testicular cancers (NONOMURA et al. 1997). In the case of embryonal tumors, LOI of *IGF2* is also linked to epigenetic silencing of the normally expressed maternal allele of *H19* (MOULTON et al. 1994; STEENMAN et al. 1994). This is consistent with a model in which these two genes are linked by competition for a shared enhancer (LEIGHTON et al. 1995). However, other tumors, including the embryonal tumor hepatoblastoma (SUGANUMA and GUPTA 1994), do not shown such linkage. Thus, the study of tumors indicates that the regulation of normal imprinting of these genes must be more complex than was initially believed. Specifically, *IGF2* can undergo relaxed imprinting, i.e., expression of the maternal allele, with no effect on *H19*. One potential mechanism for such localized alteration of *IGF2* imprinting is two regions of imprint-specific DNA methylation within the *Igf2* gene itself (SASAKI et al. 1992; FEIL et al. 1994). However, the latter observations pertain to the mouse, and a comparable region has not yet been identified in humans.

The second gene (after *H19*) that was found to undergo epigenetic silencing in Wilms' tumor was *p57^{KIP2}*. Some Wilms' tumors show apparent abnormal biallelic expression of *p57^{KIP2}*, which is due to silencing of the normally expressed maternal allele (MATSUOKA et al. 1996; THOMPSON et al. 1996). However, the silencing might not be due to altered imprinting per se but could be due to loss of some other factor necessary for normal *p57^{KIP2}* expression. This issue, that the cause of silencing of a gene may be genetically heterogeneous, applies equally well to the study of DNA methylation in cancers unrelated to imprinting. It is usually taken for granted that reduced expression of a gene linked to methylation of the promoter implies that methylation causes silencing of the gene. That is not necessarily the case, as many genes undergo silencing during development unrelated to alterations in DNA methylation (RAZIN and SZYF 1984). Furthermore, the DNA-methyltransferase-deficient knockout mouse undergoes normal silencing of a wide variety of genes during development without any change in methylation status (WALSH and BESTOR 1999). Thus, while biallelic expression of a normally imprinted gene is compelling evidence for LOI, loss of expression of a normally expressed allele, like methylation-linked silencing of a tumor-suppressor gene, must be interpreted with caution.

# 4 DNA Methylation and Loss of Imprinting

DNA methylation was first linked to loss of imprinting in cancer by studies of a CpG island within the *H19* promoter and 5′ end of the gene. Steenman et al. found that this CpG island was methylated at a level of about 50% in normal tissues (STEENMAN et al. 1994). This methylation could have been due to either heterogeneous methylation within the tissues or to imprinting, i.e., chromosome-specific methylation, with all the cells showing methylation of one parental chromosome. In order to distinguish between these two possibilities, several patients who had paternal uniparental disomy for a portion of chromosome 11 containing *H19* were identified. These patients showed methylation of the *H19* CpG island, indicating that, in normal individuals, the paternal chromosome is methylated and the maternal chromosome is not (STEENMAN et al. 1994). When Wilms' tumors were examined, it was found that the tumors showed complete methylation of this CpG island if and only if they had undergone LOI (STEENMAN et al. 1994). In addition, the normally expressed maternal allele of *H19* was silent in these patients (STEENMAN et al. 1994). Thus, the maternal chromosome in these tumors had undergone a switch in epigenotype to a paternal pattern of imprinting of: (1) *IGF2* expression; (2) *H19* expression; and (3) *H19* methylation.

However, not all tumors – not even all embryonal tumors – showed such a switch of methylation in the presence of LOI of *IGF2*. For example, hepatoblastoma was found to undergo LOI at high frequency but, in those tumors, the methylation pattern remained normal, and *H19* remained active (RAINIER et al. 1995). Thus, imprinting of *IGF2* can be controlled both locally at the gene level and also regionally by switching the parental epigenotype of this chromosomal domain.

Studies of human colorectal cancer have suggested that the P2-promoter region of *IGF2* undergoes DNA methylation as a function of aging (ISSA et al. 1996). It was suggested that this methylation might lead to activation of P1 and, therefore, biallelic expression due simply to expression of the normally non-imprinted promoter, i.e., that biallelic expression was not due to LOI at all (ISSA et al. 1996). However, direct studies of allele-specific, promoter-specific expression from our laboratory showed that is not the case and that colon cancers with biallelic expression show true LOI driven from the P2, P3, and P4 promoters, which are normally imprinted (CUI et al. 1998). Indeed, methylation of the mouse P2 promoter has not been shown to be linked to imprint-specific methylation of *Igf2* (SASAKI et al. 1992; FEIL et al. 1994). Indeed, at least in the mouse, a non-CpG-island region of DNA methylation upstream of exon 9 (3' to all of the promoters) shows allele-specific methylation (FEIL et al. 1994). This region has not yet been investigated in detail in human tumors.

Examination of LOI in colon cancer also showed that LOI was present not only in tumor tissue but also in the normal tissue of most patients with LOI in their tumor (CUI et al. 1998). Moreover, LOI was particularly common in those patients who had microsatellite instability (MSI) in their tumors (CUI et al. 1998). Most colorectal cancers with MSI do not have mutations in known mismatch-repair genes. They do, however, frequently show methylation of the *HMLH1* gene (HERMAN et al. 1998). This alteration of *HMLH1* methylation and silencing is present in the tumors but not in the normal tissue (HERMAN et al. 1998). However, LOI in these same tumors is present in normal tissue also (ISSA et al. 1996). Thus, LOI precedes the development of abnormal methylation of *hMLH1*.

What could account for such a switch in genomic imprinting, MSI in the normal tissue and the tumors, and perhaps methylation of *hMLH1*? We have hypothesized that some patients may have germline alterations in one or more genes that are not considered conventional mismatch-repairs genes but which confer a supporting function to epigenetic processing of both imprinting and DNA replication (CUI et al. 1998). The known link between normal imprinting and replication timing asynchrony (KITSBERT et al. 1993) is also consistent with this idea. By our model, maintenance of stable imprinting would be more sensitive to mutation in such genes than would maintenance of normal DNA repair, because such genes would presumably be present in the heterozygous state in normal tissue (which also shows abnormal imprinting) but would either become homozygous or exhibit additional errors in other genes in the tumors.

Finally, altered DNA methylation and genomic imprinting have been strongly linked to cancer predisposition in the hereditary disorder Beckwith-Wiedemann syndrome (BWS). As noted earlier, many Wilms' tumors show LOI of *IGF2* and altered DNA methylation of an *H19* CpG island (STEENMAN et al. 1994). The genetic disorder BWS predisposes to Wilms' and other embryonal tumors and also leads to prenatal overgrowth. BWS is associated with LOI of *IGF2* and abnormal *H19* methylation in some patients. However, until recently, the genetic etiology in most such patients remained unexplained. Recently, however, Lee et al. of our laboratory identified a CpG island in the middle of the $K_V LQT1$ gene on 11p15. This CpG island is methylated opposite to the pattern seen for *H19*; namely, it is methylated on the maternal allele and unmethylated on the paternal allele (LEE et al. 1999). Just downstream of this CpG island is a long intronic transcript that appears to be unprocessed. This transcript, termed LIT1, is expressed from the unmethylated paternal allele and silent from the methylated maternal allele, and it is transcribed through exon 10 of the $K_V LQT1$ gene, thus representing an antisense transcript (LEE et al. 1999). Thus, LIT1 shows an allele-specific pattern of expression that is the opposite of the one expressed by the sense orientation $K_V LQT1$ gene within which it resides. Lee et al. also found that most patients with BWS undergo LOI of LIT1; the normally silent maternal allele is expressed, and the normally methylated maternal allele is now unmethylated, like the paternal allele (LEE et al. 1999). Thus, LOI of LIT1 also represents a form of switching of epigenotype from maternal to paternal, in this case in a distinct genomic domain 500kb upstream of *IGF2* and *H19* (LEE et al. 1999). At the same time, Wilms' tumors do not show LOI of LIT1 (MITSUYA et al. 1999). Thus, it appears that this genomic domain is more specific to BWS, and the *IGF2/H19* domain is more specific to tumors. We hypothesize that LOI within this domain is also linked to epigenetic silencing of one or more imprinted genes within it, which include $K_V LQT1$ itself as well as $p57^{KIP2}$, which has been shown to undergo rare mutations in the coding sequence in BWS (HATADA et al. 1996; LEE et al. 1997).

While the regulation of imprinting within 11p15 may appear complex, it precisely fulfills the predictions of Haig's hypothesis of parental conflict in genomic imprinting. In this model, the paternal genome favors the growth of its particular offspring (since other offspring may arise from other fathers), while the maternally inherited genome tends to inhibit growth. Thus, a switch from maternal epigenotype to paternal epigenotype would favor growth (HAIG and GRAHAM 1991; MOORE and HAIG 1991). Furthermore, according to the model, nearby genomic regions would be in evolutionary conflict, and alterations in them could lead to distinct but overlapping phenotypes (HAIG and GRAHAM 1991; MOORE and HAIG 1991). The extreme example of a switch in epigenotype would be paternal uniparental disomy, which involves both the centromeric and telomeric imprinted domains of 11p15.

# 5  A Model of Normal Developmental Imprinting and Loss of Imprinting in Cancer

I previously suggested a model for viewing genomic imprinting within a developmental context; I believe LOI should also be viewed from the same perspective.

This model also avoids some misunderstandings that arise from a static view of the role of genomic imprinting in development or disease. The model, which is described in FEINBERG (1998), involves three steps: (1) a relatively simple modification that could involve DNA methylation, but might alternatively involve expression of a particular gene or an alteration in chromatin structure, occurs within the germline or the zygote; (2) this mark is propagated in a temporal, developmental, and physical manner along the chromosome in a manner akin to spreading X inactivation; (3) *trans*-acting factors that are developmentally regulated maintain the imprint (Fig. 1; FEINBERG 1998). Loss of imprinting in cancer should similarly be viewed from a developmental context. LOI could result from one or more of the following: deletion or chromosomal rearrangement separating the initiating center from the imprinted target gene; alterations in local control elements that maintain epigenetic silencing; or loss of the expression of *trans*-acting factors necessary to maintain a stable imprint (Fig. 1; FEINBERG 1998).

Where might DNA methylation fit into this developmental model of genomic imprinting? At all three levels described. First, the initial mark could be due to DNA methylation itself. We suspect that such a mechanism may be involved in LOI of LIT1 in BWS associated with loss of methylation of a CpG island that is normally methylated on the silent maternal allele (LEE et al. 1999). However, we do not yet have proof that LOI of LIT1 leads to alterations in allele-specific expression of other genes within this domain. Nevertheless, this hypothesis is consistent with our previous observation that all of the balanced chromosomal-rearrange-

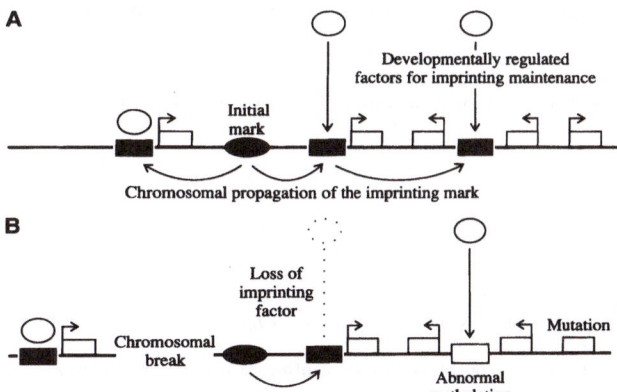

**Fig. 1.** Imprinting as a dynamic process in normal development that is disrupted in cancer. **A** As described, this process involves three steps. (1) A relatively simple modification occurs within the germline or the zygote; this could involve DNA methylation but might, alternatively, involve expression of a particular gene or an alteration in chromatin structure. (2) This mark is propagated in a temporal, developmental, and physical manner along the chromosome in a manner akin to spreading X inactivation. (3) *Trans*-acting factors that are developmentally regulated maintain the imprint. **B** Loss of imprinting in cancer should similarly be viewed from a developmental context, and it could result from one or more of the following: either deletion or chromosomal rearrangement separating the initiating center from the imprinted target gene; alterations in local control elements that maintain epigenetic silencing; or loss of the expression of *trans*-acting factors necessary to maintain a stable imprint. Reprinted from Feinberg (1998) with permission

ment breakpoints within the BWS cluster region (BWSCR1) interrupt the $K_VLQT1$ gene (HOOVERS et al. 1995). DNA methylation is also likely to play a role in the maintenance of the local control of imprinting. For example, we and others have previously shown that altered DNA methylation is linked to loss of expression of the maternal *H19* allele and activation of the maternal *IGF2* allele in Wilms' tumor (MOULTON et al. 1994; STEENMAN et al. 1994). This idea is also consistent with the observation that loss of DNA methyltransferase leads to loss of imprinting of many (but not all) imprinted genes (LI et al. 1993; CASPARY et al. 1998). Finally, DNA methylation may be involved in the regulation of *trans*-acting factors that modulate genomic imprinting. While there is no evidence for such a role yet, it is possible that experiments in which 5-aza-2'deoxycytidine are used to re-establish a normal pattern of imprinting in tumor cells may actually have an effect on a modifier gene whose protein product is necessary to maintain a normal pattern of imprinting but is not responsible for the imprinting mark itself (BARLETTA et al. 1997).

What kind of gene could this be? A tantalizing clue may come from the recent discovery that altered DNA methylation in the flowering plant *Arabidopsis thaliana* can be due (trivially) to altered DNA methyltransferase or, importantly, to mutations in a *SWI2/SNF2*-like gene (JEDDELOH et al. 1999). The implication is that chromatin-remodeling proteins that are known to be involved across diverse species, from yeast to plants to animals, may serve a specialized role in chromatin remodeling, a role specific to gene silencing. Among the most exciting frontiers in the study of genomic imprinting in development and disease is the identification of such *trans*-acting factors and their role in controlling expression over multi-gene-imprinted domains.

*Acknowledgement.* This work was supported by National Institutes of Health Grant CA65145.

# References

Ainscough JF, Koide T, Tada M, Barton S, Surani MA (1997) Imprinting of *Igf2* and *H19* from a 130-kb *YAC* transgene. Develop 124:3621–3632

Barletta JM, Rainier S, Feinberg AP (1997) Reversal of loss of imprinting in tumor cells by 5-aza-2'-deoxycytidine. Cancer Res 57:48–50

Bartolomei M, Zemel S, Tilghman SM (1991) Parental imprinting of the mouse *H19* gene. Nature 351:153–155

Bartolomei MS, Webber AL, Brunkow ME, Tilghman SM (1993) Epigenetic mechanisms underlying the imprinting of the mouse *H19* gene. Genes Develop 7:1663–1673

Birger Y, Shemer R, Perk J, Razin A (1999) The imprinting box of the mouse *Igf2r* gene. Nature 397: 84–88

Buiting K, Dittrich B, Robinson WP, Guitart M, Abeliovich D, Lerer I, Horsthemke B (1994) Detection of aberrant DNA methylation in unique Prader-Willi syndrome patients and its diagnostic implications. Hum Mol Genet 3:893–895

Buiting K, Saitoh S, Gross S, Dittrich B, Schwartz S, Nicholls RD, Horsthemke B (1995) Inherited microdeletions in the Angelman and Prader-Willi syndromes define an imprinting centre on human chromosome 15. Nat Genet 9:395–400

Caspary T, Cleary MA, Baker CC, Guan XJ, Tilghman SM (1998) Multiple mechanisms regulate imprinting of the mouse distal chromosome 7 gene cluster. Molecular and Cellular Biology 18: 3466–3474

Cattanach BM, Beechey CV (1990) Autosomal and X-chromosome imprinting. Develop Suppl:63–72

Crouse HV (1960) The controlling element in sex chromosome behavior in *Sciara*. Genetics 45:1425–1443

Cui H, Horon IL, Ohlsson R, Hamilton SR, Feinberg AP (1998) Loss of Imprinting in normal tissue of colorectal cancer patients with microsatellite instability. Nat Med 4:1276–1280

DeChiara TM, Robertson EJ, Efstratiadis A (1991) Parental imprinting of the mouse insulin-like growth factor-2 gene. Cell 64:849–859

Douc-Rasy S, Barrois M, Fogel S, Ahomadegbe JC, Stehelin D, Coll J, Riou G (1996) High incidence of loss of heterozygosity and abnormal imprinting of *H19* and *IGF2* genes in invasive cervical carcinomas. Uncoupling of *H19* and *IGF2* expression and biallelic hypomethylation of *H19*. Oncogene 12:423–430

Feil R, Walter J, Allen ND, Reik W (1994) Developmental control of allelic methylation in the imprinted mouse *IGF2* and *H19* genes. Develop 120:2933–2943

Feinberg AP (1993) Genomic imprinting and gene activation in cancer. Nat Genet 4:110–113

Feinberg AP (1998) Genomic imprinting as a developmental process disturbed in cancer. In: Reik W, Surani A (eds) Genomic imprinting: frontiers in molecular biology. IRL Press: Oxford

Feinberg AP (1999) Imprinting of a genomic domain of 11p15 and loss of imprinting in cancer: an introduction. Cancer Res 59:1743s–1746s

Feinberg AP, Vogelstein B (1983) Hypomethylation distinguishes genes of some human cancers from their normal counterparts. Nature 301:89–92

Ferguson-Smith AC, Sasaki H, Cattanach BM, Surani MA (1993) Parental-origin-specific epigenetic modification of the mouse *H19* gene. Nature 362:751–755

Haig D, Graham C (1991) Genomic imprinting and the strange case of the insulin-like growth-factor-II receptor. Cell 64:1045–1046

Hatada H, Ohashi Y, Fukushima Y, Kaneko M, Inoue Y, Komoto A, Okada S, Ohishi A, Nabetani H, Morisaki M, Nakayama N, Niikawa N, Mukai T (1996) An imprinted gene $p57^{KIP2}$ is mutated in Beckwith-Wiedemann syndrome. Nat Genet 14:171–173

Herman JG, Umar A, Polyak K, Graff JR, Ahuja N, Issa JP, Markowitz S, Willson JK, Hamilton SR, Kinzler KW, Kane MF, Kolodner RD, Vogelstein B, Kunkel TA, Baylin SB (1998) Incidence and functional consequences of *hMLH1* promoter hypermethylation in colorectal carcinoma. Proc Natl Acad Sci USA 95:6870–6875

Hoovers JMN, Kalikin LM, Johnson LA, Alders M, Redeker B, Law DJ, Bliek J, Steenman M, Benedict M, Wiegant J, Lengauer C, Taillon-Miller P, Schlessinger D, Edwards MC, Elledge SJ, Ivens A, Westerveld A, Little P, Mannens M, Feinberg AP (1995) Multiple genetic loci within 11p15 defined by Beckwith-Wiedemann syndrome rearrangement breakpoints and subchromosomal transferable fragments. Proc Natl Acad Sci USA 92:12456–12460

Hu R-J, Lee MP, Connors TD, Johnson LA, Burn TC, Su K, Landes GM, Feinberg AP (1997) A 2.5-Mb transcript map of a tumor-suppressing sub-chromosomal transferable fragment from 11p15.5, and isolation and sequence analysis of three novel genes. Genomics 46:9–17

Issa JP, Vertino PM, Boehm CD, Newsham IF, Baylin SB (1996) Switch from monoallelic to biallelic human IGF2 promoter methylation during aging and carcinogenesis. Proc Natl Acad Sci USA 93:11757–11762

Jeddeloh JA, Stokes TL, Richards EJ (1999) Maintenance of genomic methylation requires a SWI2/SNF2-like protein. Nat Genet 22:94–97

Kajii T, Ohama K (1977) Androgenetic origin of hydatidiform mole. Nature 268:633

Kitsbert D, Selig S, Brandeis M, Simon I, Keshet I, Driscoll DJ, Nicholls RD, Cedar H (1993) Allele-specific replication timing of imprinted gene regions. Nature 364:459–463

Ledbetter DH, Engel E (1995) Uniparental disomy in humans: Development of an imprinting map and its implications for prenatal diagnosis. Hum Mol Genet 4:1757–1764

Lee MP, DeBaun M, Randhawa GS, Reichard BA, Feinberg AP (1997) Low frequency of $p57^{KIP2}$ mutation in Beckwith-Wiedemann syndrome. Am J Hum Genet 61:304–309

Lee MP, DeBaun MR, Mitsuya K, Galonek HL, Brandenburg S, Oshimura M, Feinberg AP (1999) Loss of imprinting of a paternally expressed transcript, with antisense orientation to *KVLQT1*, occurs frequently in Beckwith-Wiedemann syndrome and is independent of insulin-like growth-factor-II imprinting. Proc Natl Acad Sci USA 96:5203–5208

Leighton PA, Ingram RS, Eggenschwiler J, Efstratladis A, Tilghman SM (1995) Disruption of imprinting caused by deletion of the *H19* gene region in mice. Nature 375:34–39

Li E, Beard C, Jaenisch R (1993) Role for DNA methylation in genomic imprinting. Nature 366:362–365

Linder D, McCaw B, Kaiser X, Hecht F (1975) Parthenogenetic origin of benign ovarian teratomas. N Engl J Med 292:63–66

Matsuoka S, Thompson JS, Edwards MC, Barletta JM, Grundy P, Kalikin LM, Harper JW, Elledge SJ, Feinberg AP (1996) Imprinting of the gene encoding a human cyclin-dependent kinase inhibitor, $p57^{KIP2}$, on chromosome 11p15. Proc Natl Acad Sci USA 93:3026–3030

Mitsuya K, Meguro M, Lee MP, Katoh M, Schulz TC, Kugoh H, Yoshida MA, Niikawa N, Feinberg AP, Oshimura M (1999) LIT1, an imprinted antisense RNA in the human $K_V$LQT1 locus identified by screening for differentially expressed transcripts using monochromosomal hybrids. Hum Mol Genet, in press

Moore T, Haig D (1991) Genomic imprinting in mammalian development: a parental tug-of-war. Trends Genet 7:45–49

Morison IM, Reeve AE (1998) A catalogue of imprinted genes and parent-of-origin effects in humans and animals. Hum Mol Genet 7:1599–1609

Moulton T, Crenshaw T, Hao Y, Moosikasuwan J, Lin N, Dembitzer F, Hensle T, Weiss L, McMorrow L, Loew T, Kraus W, Gerald W, Tycko B (1994) Epigenetic lesions at the *H19* locus in Wilms' tumour patients. Nat Genet 7:440–447

Nonomura N, Miki T, Nishimura K, Kanno N, Kojima Y, Okuyama A (1997) Altered imprinting of the *H19* and insulin-like growth-factor-II genes in testicular tumors. J Urol 157:1977–1979

Ogawa O, Eccles MR, Szeto J, McNoe LA, Yun K, Maw MA, Smith PJ, Reeve AE (1993) Relaxation of insulin-like growth-factor-II gene imprinting implicated in Wilms' tumour. Nature 362:749–751

O'Keefe D, Dao D, Zhao L, Sanderson R, Warburton D, Weiss L, Anyane-Yeboa K, Tycko B (1997) Coding mutations in *p57KIP2* are present in some cases of Beckwith-Wiedemann syndrome but are rare or absent in Wilms tumors. Am J Hum Genet 61:295–303

Plass C, Shibata H, Kalcheva I, Mullins L, Kotelevtseva N, Mullins J, Kato R, Sasaki H, Hirotsune S, Okazaki Y, Held WA, Hayashizaki Y, Chapman VM (1996) Identification of *Grf1* on mouse chromosome 9 as an imprinted gene by RLGS-M. Nat Genet 14:106

Rainier S, Dobry CJ, Feinberg AP (1995) Loss of imprinting in hepatoblastoma. Cancer Res 55:1836–1838

Rainier S, Johnson LA, Dobry CJ, Ping AJ, Grundy PE, Feinberg AP (1993) Relaxation of imprinted genes in human cancer. Nature 362:747–749

Razin A, Szyf M (1984) DNA methylation patterns: formation and function. Biochim Biophys Acta 782:331–342

Reik W, Collick A, Norris ML, Barton SC, Surani A (1987) Genomic imprinting determines methylation of parental alleles in transgenic mice. Nature 328:248–251

Reis A, Dittrich B, Greger V, Buiting K, Lalande M, Gillessen-Kaesbach G, Anvret M, Horsthemke B (1994) Imprinting mutations suggested by abnormal DNA methylation patterns in familial Angelman and Prader-Willi syndromes. Am J Hum Genet 54:741–747

Riesewijk AM, Hu L, Schulz U, Tariverdian G, Hoglund P, Kere J, Ropers H-H, Kalscheuer VM (1997) Monoallelic expression of human *PEG1/MEST* is paralleled by parent-specific methylation in fetuses. Genomics 42:236–244

Sapienza C (1990) Genome imprinting, cellular mosaicism and carcinogenesis. Mol Carcinog 3:118–121

Sapienza C, Peterson AC, Rossant J, Balling R (1987) Degree of methylation of transgenes is dependent on gamete of origin. Nature 328:251–254

Sasaki H, Jones PA, Chaillet JR, Ferguson-Smith AC, Barton SC, Reik W, Surani MA (1992) Parental imprinting: potentially active chromatin of the repressed maternal allele of the mouse insulin-like growth factor II (*IGF2*) gene. Genes Develop 6:1843–1856

Scrable H, Cavenee W, Ghavimi F, Lovell M, Morgan K, Sapienza C (1989) A model for embryonal rhabdomyosarcoma tumorigenesis that involves genome imprinting. Proc Natl Acad Sci USA 86:7480–7484

Shibata H, Yoshino K, Sunahara S, Gondo Y, Katsuki M, Ueda T, Kamiya M, Muramatsu M, Murakami Y, Kalcheva I, Plass C, Chapman VM, Hayashizaki Y (1996) Inactive allele-specific methylation and chromatin structure of the imprinted gene *U2af1-rs1* on mouse chromosome 11. Genomics 35:248–252

Steenman MJC, Rainier S, Dobry CJ, Grundy P, Horon IL, Feinberg AP (1994) Loss of imprinting of *IGF2* is linked to reduced expression and abnormal methylation of *H19* in Wilms' tumor. Nat Genet 7:433–439

Stoger R, Kubicka P, Liu C-G, Kafri T, Razin A, Cedar H, Barlow DP (1993) Maternal-specific methylation of the imprinted mouse *IGF2r* locus identifies the expressed locus as carrying the imprinting signal. Cell 738:61–71

Swain JL, Stewart TA, Leder P (1987) Parental legacy determines methylation and expression of an autosomal transgene: a molecular mechanism for parental imprinting. Cell 50:719–727

Szabo PE, Mann JR (1995) Allele-specific expression and total expression levels of imprinted genes during early mouse development: implications for imprinting mechanisms. Genes Develop 9:3097–3108

Takeda S, Kondo M, Kumada T, Koshikawa T, Ueda R, Nishio M, Osada H, Suzuki H, Nagatake M, Washimi O, Takagi K, Takahashi T, Nakao A (1996) Allelic-expression imbalance of the insulin-like growth factor 2 gene in hepatocellular carcinoma and underlying disease. Oncogene 12:1589–1592

Thompson JS, Reese KJ, DeBaun MR, Perlman EJ, Feinberg AP (1996) Reduced expression of the cyclin-dependent kinase inhibitor $p57^{KIP2}$ in Wilms tumor. Cancer Res 56:5723–5727

Thorvaldsen JL, Duran KL, Bartolomei MS (1998) Deletion of the *H19* differentially methylated domain results in loss of imprinted expression of *H19* and *Igf2*. Genes Dev 12:3693–3702

Tremblay KD, Saam JR, Ingram RS, Tilghman SM, Bartolomei MS (1995) A paternal-specific methylation imprint marks the alleles of the mouse *H19* gene. Nat Genet 9:407–413

Uyeno S, Aoki Y, Nata M, Sagisaka K, Kayama T, Yoshimoto T, Ono T (1996) *IGF2* but not *H19* shows loss of imprinting in human glioma. Cancer Res 56:5356–5359

Walsh CP, Bestor TH (1999) Cytosine methylation and mammalian development. Genes Dev 13:26–34

Wutz A, Smrzka OW, Schweifer N, Schellander K, Wagner EF, Barlow DP (1997) Imprinted expression of the *Igf2r* gene depends on an intronic CpG island. Nature 389:745–749

Zhan S, Shapiro DN, Helman LJ (1994) Activation of an imprinted allele of the insulin-like growth-factor-II gene implicated in rhabdomyosarcoma. J Clin Invest 94:445–448

# CpG-Island Methylation in Aging and Cancer

J.-P. Issa

# 1 Introduction

CpG islands are small (0.5–2kb) CpG-rich areas that are commonly found in the promoters of expressed genes (BIRD 1986). In the early studies that established the concept of CpG islands, it became clear that these CpG-rich areas are devoid of cytosine DNA methylation regardless of the expression status of the gene. In fact, this lack of methylation was incorporated in the early definitions of CpG islands (BIRD 1986). In the past few years, hypermethylation of some promoter-associated CpG islands have been described in various malignancies in association with silencing of the gene (BAYLIN et al. 1998; JONES and LAIRD 1999). In a search for

The Johns Hopkins Oncology Center, 424 N. Bond Street, Baltimore, MD 21231, USA
E-mail: jpissa@jhmi.edu

potential causes of this phenomenon, it has become apparent that, in many cases, aberrant methylation precedes full-blown malignancy and can often be detected in histologically normal tissues. This fact appears to contradict the original findings of lack of CpG-island methylation in normal tissues. A resolution of this conflict came from the realization that CpG-island methylation in normal tissues may, in itself, be a pathologic event that can be related to age or time-dependent events (ISSA et al. 1994). In this chapter, our current understanding of age-related methylation changes will be reviewed along with a discussion of their potential causes and functional consequences. In particular, a model will be proposed whereby promoter methylation and gene silencing provide a mechanistic link between the processes of aging and neoplasia.

## 2 Early Studies of DNA-Methylation Changes in Aging Tissues

### 2.1 Genomic Hypomethylation

The identification of CpG methylation as a potential mechanism of epigenetic inheritance has lead to speculations that it might be involved in the aging process (HOLLIDAY 1987). Early studies of this phenomenon focussed on measuring total 5-methylcytosine (5-mC) levels in cells aged in vitro and in tissues in vivo. Wilson et al. reported progressive losses of 5-mC in normal fibroblasts during in vitro passaging (WILSON and JONES 1983a). These studies were confirmed by others (HOLLIDAY 1985). Studying the same process, Holliday and colleagues found that treatment with the methylation inhibitor 5-azacytidine shortens the lifespan of cultured fibroblasts (HOLLIDAY 1986), possibly suggesting a direct link between genomic hypomethylation and the aging process. 5-Azacytidine, however, is a considerably toxic drug, and may have induced effects independent of loss of 5-mC. Genomic hypomethylation has also been observed in aged tissues in vivo (MAYS-HOOPES et al. 1986; WILSON et al. 1987; DRINKWATER et al. 1989).

The majority of methylated cytosines are located within repeated elements, including satellite DNA and Alu repeats (BESTOR and TYCKO 1996). Therefore, it was not surprising to find that much of this age-related demethylation affected repeated sequences (ROMANOV and VANYUSHIN 1981; MAYS-HOOPES et al. 1986). Gene-specific hypomethylation has also been described in some cases, and can affect oncogenes such as c-MYC (ONO et al. 1986). In these cases, loss of methylation involves cytosines that are in the coding sequences and introns and are relatively distant from the promoter and 5′ regulatory regions of the genes.

These early studies led to the speculation that hypomethylation may partially explain the changes in gene expression observed in aging cells and tissues (HOLLIDAY 1987). Central to this hypothesis was the emerging relationship between cytosine methylation and inhibited gene expression. More recent data, however, have raised questions regarding the validity of a relationship between DNA methylation outside of gene promoters and transcriptional repression of the involved gene. Indeed, several genes are transcribed despite fairly heavy methylation within their coding sequences (TORNALETTI and PFEIFER 1995; JONES and LAIRD 1999), and mice deficient in Dnmt1 do not appear to activate genes other than those regulated by promoter CpG-island methylation (WALSH and BESTOR 1999). The functional significance of age-related hypomethylation, therefore, remains to be determined. Hypomethylation has also been linked to chromosomal instability (CHEN et al. 1998), although it is unclear whether the degree of age-related hypomethylation observed could have a similar effect. One function of DNA methylation that has recently been championed is in the transcriptional suppression of potential expressed retroviral elements in the genome (BESTOR and TYCKO 1996). Because age-related methylation primarily affects such repeated sequences, it is possible that increased expression of Alus and other parasitic DNA will be found in aging cells.

Despite having been described almost two decades ago, the causes of age-related hypomethylation remain obscure. Genomic losses of 5-mC follow carcinogen exposure (WILSON and JONES 1983b), and it is possible that the age effects are due to cumulative exposure to exogenous mutagens or endogenous reactive oxygen species. This hypothesis awaits confirmation but could potentially be tested in animal models. However, marked tissue-specific differences in genomic methylation have been noted in vivo (GAMA-SOSA et al. 1983), and some studies suggest that hypomethylation may, in fact, be a feature of proliferating cells (HOAL-VAN HELDEN and VAN HELDEN 1989; GOODMAN and COUNTS 1993). Increased proliferation during the aging process has been noted in some tissues in vivo (RONCUCCI et al. 1988), possibly providing a partial explanation for this phenomenon.

## 2.2 Gene-Specific Hypermethylation

While most early investigations focussed on age-related hypomethylation, a few studies reported on actual hypermethylation at selected loci. In particular, a series of reports from Ono et al. described progressive, age-related hypermethylation of the c-MYC and c-FOS genes in the liver and other tissues of both humans and rodents (CHOI et al. 1996; ONO et al. 1993). For both genes, a large region flanking the transcription-start area remained unmethylated, but the borders of this region showed progressive spreading of methylation towards the exon-1/intron-1 region. While the CpG islands of both genes were unaffected by age-related methylation, the absolute area of protection from methylation became progressively smaller with age. Interestingly, in mice, these gene-specific hypermethylation events with age occurred in the face of simultaneous global losses in 5-mC content. Furthermore, c-MYC hypermethylation could be attenuated by caloric restriction in this model (MIYAMURA et al. 1993).

Post-embryonic development is also associated with changes in methylation at multiple loci (RAZIN and SHEMER 1995). These changes are tissue-specific and include both hypermethylation and hypomethylation. Examples of development-related hypermethylation include exon-1 sequences in the mouse MyoD gene (GHAZI et al. 1992) and coding sequences in the *ApoAI* gene (SHEMER et al. 1991), where a CpG island is actually involved in this process.

The contribution of development and age-related methylation to changes in expression of the affected genes is unclear. In some cases, hypermethylation is accompanied by decreased expression of the gene (ONO et al. 1989), but this is not a universal finding. In fact, because the changed methylation occurs outside the promoters of the genes, it is unclear whether these events cause or simply follow changes in gene expression. As mentioned previously, it has been argued that methylation of cytosine residues outside of promoter-associated CpG islands does not, in fact, have a role in the regulation of gene expression (except for Alus and other repetitive elements). However, while the functional significance of the above-described alteration is unknown, it clearly established the fact that gene-specific methylation patterns do have a significant degree of plasticity and that adult cells are indeed capable of de novo methylation. Such plasticity in methylation patterns has also been observed at X-chromosome loci, and these data have challenged the traditional models of DNA-methylation formation (RIGGS et al. 1998). In particular, the model that restricted de novo methylation to embryogenesis with a strictly maintenance methylase activity in adult cells cannot account for the changes observed during development and aging.

# 3 CpG-Island Methylation and Aging

While the function and mediators of DNA methylation were being elucidated, studies of CpG-island methylation in cancer provided an added dimension to the equation. CpG islands are short, CpG-rich areas that are normally unmethylated and are present in the promoter region of about half of all human genes (BIRD 1986). In the past decade, promoter CpG-island methylation has been demonstrated to be an essential part of gene silencing on the inactive X-chromosome (GOTO and MONK 1998) and at selected autosomal genes that display parent allelic expression patterns or imprinting (BARLOW 1995). Unlike methylation in coding regions, promoter CpG-island methylation is strongly linked to silencing of transcription, probably mediated by binding of methylated DNA-binding proteins and the recruitment of a protein complex that results in a closed chromatin structure (KASS et al. 1997). Considerable interest in this mechanism of epigenetic silencing was generated recently by a series of reports that linked aberrant CpG-island methylation with tumor-suppressor-gene silencing in neoplasia (BAYLIN et al. 1998; JONES and LAIRD 1999). As discussed in other chapters in this book, cancers appear to thwart this selective mechanism of permanent gene silencing and use it to suppress the expression and function of multiple genes. While this process was initially thought to be specific to the cancer process, studies involving methylation of the estrogen receptor (ER) gene in the colon revealed an unusual pattern of partial CpG-island methylation in normal tissues; this pattern was linked to the aging process (Issa et al. 1994). Further studies revealed that age-related methylation is, in fact, a common process that affects multiple genes (Ahuja et al. 1998) and may be part of the long-sought mechanism linking aging and neoplasia.

## 3.1 Age-Related Methylation of the *ER* Gene

One-third of human breast cancers do not express functional ERs, despite having no apparent structural defects in the *ER* gene. *ER* has a typical CpG island in its 5' region, which is hypermethylated in *ER*-negative breast cancer (Ottaviano et al. 1994). Pharmacologic demethylation results in re-expression of functional ERs, suggesting that hypermethylation is critical to maintaining the silenced state of this gene in breast cancer (Ferguson et al. 1995). Similar to breast cancer, the incidence of colon cancer in women is modulated by reproductive factors, such as number of pregnancies and post-menopausal hormonal replacement (Potter et al. 1993). This suggested a potential contribution of the *ER* system to the development of colon cancer as well. Indeed, all colon-cancer cell lines and primary tumors studied (both benign, adenomatous polyps and cancers) were hypermethylated at the *ER*-gene locus (Issa et al. 1994). As in breast cancer, this methylation is associated with absent transcription of the *ER* gene in colonic tumors. Forced re-expression of *ER* by transfection of an *ER*-expression construct into a colonic cell line resulted in marked reduction in growth and clonogenicity. These data suggested that *ER* methylation might be an important step in the pathophysiology of colonic tumors.

Upon studying the *ER* CpG island in colorectal tumors, it became apparent that partial methylation can be seen in the normal colon of patients with cancer, in addition to controls (examples in Fig. 1). Upon further investigation, this methylation was determined to be an age-related event, being almost undetectable in young individuals and becoming progressively more prominent with age (Issa et al. 1994). In a study of almost 400 patients aged 8–90 years, *ER* methylation was found to increase linearly with age at a rate of 0.2–0.3% per year (Fig. 1). This study included over 300 patients without colonic tumors, suggesting that this defect is not limited to patients affected with the disease (Issa et al., in preparation). Interestingly, even though the age-related increase in *ER* methylation appears to be fairly uniform in the general population, there were differences as large as tenfold in the extent of methylation within each age group studied. These marked individual differences in the extent of *ER* methylation suggest an influence of either genetic factors or environmental exposures (or both) on age-related methylation. Nevertheless, because all colonic tumors examined (including small adenomatous polyps) showed this defect, it was hypothesized that the cell of origin of these tumors was one where the *ER* gene was hypermethylated.

In addition to individual variation, there are marked, tissue-specific differences in age-related *ER*-gene methylation. In the colon, *ER* methylation is predominantly a phenomenon of the epithelium, where methylation ranged between 30% and 70% in four patients, while methylation in the stroma ranged from 5–10% in the same patients (Ahuja et al. 1998). Furthermore, *ER* methylation is about twofold higher

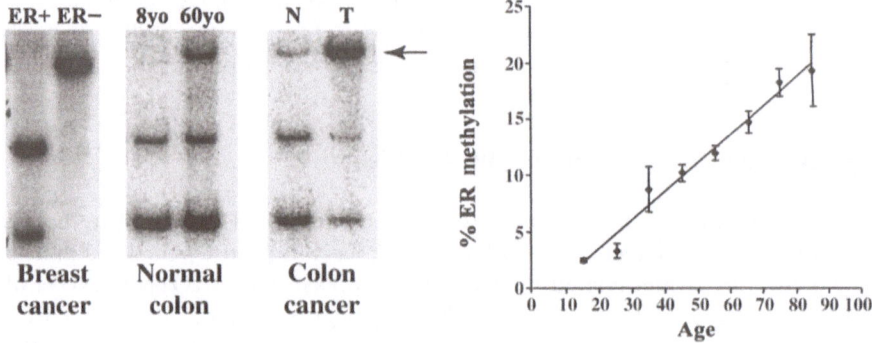

**Fig. 1.** Estrogen-receptor gene (*ER*) methylation in cancer and aging colon. Shown on the *left* are Southern blots probed with an *ER* exon-1 probe. The *arrow* points to the band that indicates CpG-island methylation. Note that methylation is absent in an *ER*+ breast-cancer cell line and is complete in an *ER*− breast cancer cell line (*left*). In the colon (*right*), *ER* is slightly methylated in normal colon (*N*) and highly methylated in colon cancer (*C*). The middle Southern blot shows that methylation is undetectable in the normal colon of an 8-year-old child but is easily detectable in the normal colon of a 60-year-old individual. Shown on the *right* is a graph correlating *ER* methylation in normal colon with age. Methylation was measured by Southern blotting and densitometry for 400 individuals and was averaged for each decade from the second to ninth. Each *point* represents the mean for each decade, and the *error bars* represent the standard error of the mean. The data was pooled from several different studies (Issa et al. 1994; Issa et al., in preparation)

in the sigmoid colon than in the cecum, suggesting that these two anatomic locations of the colon behave differently with respect to age-related methylation (Issa et al. 1994). *ER* methylation is lowest (and barely detectable) in breast epithelium, bronchial epithelium and endometrium, intermediate in lung stroma, kidney, prostate, peripheral blood cells and bone marrow, and highest in the liver, an organ that has been reported to express the *ER* gene (Ottaviano et al. 1994; Issa et al. 1996a, 1996c; Ahuja et al. 1998; Issa et al., unpublished data). These studies indicate that age-related methylation at the *ER* locus is a highly tissue-specific process and may well be modulated by expression of the *ER* gene, being lowest in tissues with high expression. In addition, the high levels of *ER* methylation in colonic epithelium and liver may be related to exposure to environmental and/or dietary carcinogens. Indeed, in rodent models of lung adenocarcinomas, the degree of *ER* methylation in the tumors is dependent on the type of carcinogenic insult that induced tumor formation (Issa et al. 1996a).

*ER* methylation is not a unique property of tissues predisposed to neoplasia. Indeed, it may also be involved in non-neoplastic age-related diseases. Atherosclerosis, coronary artery disease (CAD) and heart failure, for example, are predominantly diseases of older individuals. Women have a delayed onset of CAD when compared with men, and circulating estrogens have been shown to protect women against the development of atherosclerosis (Nathan and Chaudhuri 1997). The *ER* gene is expressed in cardiovascular tissues and may mediate some of the beneficial effects of circulating estrogens on CAD (Mendelsohn and Karas 1994). Interestingly, it has been demonstrated by immunohistochemistry that women with CAD had markedly diminished expression of the *ER* gene in their aorta when compared with age-matched controls, and it appeared that older women uniformly had fewer ERs in their aorta than younger women (LOSORDO et al. 1994). These observations prompted studies of *ER*-gene methylation in cardiovascular tissues (POST et al. 1999). In heart muscle, *ER* methylation ranged from 6.1% to 17.5%, and there was an excellent correlation between age and *ER* methylation ($R = 0.59$, $P = 0.01$). In healthy-appearing arteries, there was very little detectable methylation, similar to what was observed in breast and bronchial epithelium. In sharp contrast, several atherectomy samples from patients of comparable age had easily detectable *ER* methylation and, on average, diseased arteries had significantly more *ER* methylation than normal-appearing arteries. Remarkably, this methylation probably comes from proliferating smooth-muscle cells (which are part of the pathogenesis of the disease), because they have high levels of *ER* methylation in cell culture. These data suggest that progressive CpG-island methylation may also be a feature of age-related diseases other than neoplasia.

## 3.2 Age-Related Methylation of Other Genes

### 3.2.1 Insulin-Like-Growth-Factor-II Gene

The insulin-like-growth-factor-II gene (IGF2) is a fetal growth promoter imprinted in rodents and in humans, being expressed exclusively from the paternal allele (GIANNOUKAKIS et al. 1993). In humans, the IGF2 gene has four promoters (P1–P4), with four different leader exons (SUSSENBACH et al. 1993). P2–P4, which are relatively close to each other, are contained in a CpG island and are coordinately imprinted. By contrast, P1, which is located more than 20kb upstream of P2, is not contained in a CpG island and is expressed bi-allelically in many human tissues, suggesting that it escapes imprinting (VU and HOFFMAN 1994). Southern-blot and polymerase chain reaction (PCR) analysis revealed that the P2–P4 IGF2 CpG island is methylated only on the silenced maternal allele. Strikingly, P2–P4 methylation increases progressively with age and, in the colon, this increased methylation can be accounted for, in part, by de novo methylation of the previously unmethylated paternal allele (ISSA et al. 1996b). Most neoplasms studied had even more extensive P2–P4 methylation and, in tumor cell lines, hypermethylation at P2–P4 resulted in a marked reduction of IGF2 transcription from P3. By contrast, P1, the upstream promoter that is not contained in a CpG island, was expressed in most cell lines examined.

These findings may be particularly important because they provide a potential clue to the mechanism of age-related methylation. The data indicate that areas predisposed to allele-specific methylation may be particularly at risk for this phenomenon. Thus, the local signals that initiate allele silencing may be similar to the signals that direct age-related hypermethylation. In addition, the finding of methylation within the previously unmethylated allele suggests that this gene is predisposed to age-related methylation regardless of expression, either because of some local DNA sequence that directs de novo methylation with age or, alternatively, through "spreading" of methylation from the methylated allele to the opposite allele during aging. Previously, spreading of methylation across alleles was suggested to account for a phenomenon termed co-suppression in plants, which is a process whereby a transgene that has become methylated and inactivated results in age-dependent methylation and inactivation of the homologous endogenous gene (FLAVELL 1994). To explain this phenomenon, it has been proposed that the methylated transgene and the endogenous gene pair transiently by homologous recombination and that DNA-methyltransferase recognizes the paired strands as hemi-methylated DNA and "spread" methylation from the transgene to the endogenous gene (MALOISEL and ROSSIGNOL 1998). Such homologous pairing has been observed at imprinted loci in humans (LASALLE and LALANDE 1996). Interestingly, homology-dependent spreading of methylation with age may not always require prior methylation of one allele. Many human genes have pseudogenes, which are inactive and usually hypermethylated. If these pseudogenes share significant homology with the 5′ region of an expressed gene, one can envision homologous pairing and spreading of methylation during the aging process.

### 3.2.2 *MYOD1*

*MYOD1* is a muscle-specific transcription factor that was cloned, based in part on reactivation using methylation inhibitors (DAVIS et al. 1987). In fact, the CpG island of *MYOD1* is heavily methylated in a variety of cell lines and primary tumors (RIDEOUT et al. 1994). In the colon, methylation of the *MYOD1* CpG island is nearly universal in tumors, and it can easily be detected in adjacent, normal-appearing mucosa (AHUJA et al. 1998). Just like the *ER* gene, *MYOD1* methylation increases progressively with age. Interestingly, the most CpG-dense area of the island actually corresponds to exon 2 of the gene, and age-related methylation is more prominent there than in the area immediately upstream of exon 1. It is possible that methylation actually begins in the coding region of the gene and progressively spreads upstream (and presumably downstream as well), to eventually involve the promoter of the gene per se. *MYOD1* is not expressed in adult tissues other than skeletal muscle.

### 3.2.3 *N33*

*N33* is a gene on 8p that maps to an area of frequent loss of heterozygosity in prostate cancer (MACGROGAN et al. 1996). *N33* is presumed to function as an oligosaccharyl transferase and does not appear to be the targeted tumor-suppressor gene in this area. Nevertheless, this gene was reported to be frequently methylated in colorectal cancer, and Southern-blot analysis demonstrated it to be affected by age-related methylation in normal colon (AHUJA et al. 1998). Interestingly, the gene is methylated in 70% of glioblastoma multiforme and is much more frequently affected in tumors of older patients than in those of younger ones (LI et al. 1998).

### 3.2.4 *HIC1*

*HIC1*, a candidate tumor-suppressor gene, was cloned from an area of frequent loss of heterozygosity on 17p13.3 that was found to be frequently hypermethylated in multiple human tumors (WALES et al. 1995). *HIC1* methylation, however, can be detected in several normal tissues, including breast epithelium (FUJII et al. 1998), brain (LI et al. 1998), prostate (MORTON et al. 1996), kidney (MAKOS et al. 1993) and liver. In the prostate, *HIC1* methylation progressively increases with age (Issa et al., unpublished) and is very frequently seen in malignant neoplasms. By contrast, *HIC1* is not methylated in normal colons, including colons of older individuals (AHUJA et al. 1998).

### 3.2.5 *Versican* and *PAX6*

Both *Versican* and *PAX6* were isolated from a colorectal-cancer cell line using a novel technique developed to clone differentially methylated CpG islands (TOYOTA et al. 1999b). *Versican* is a secreted proteoglycan that is upregulated by overexpression of wild-type *RB1* (ROHDE et al. 1996). It is expressed at relatively high levels in normal colon and is frequently methylated in neoplastic cell lines and primary tumors. In normal colon, methylation of the *Versican* promoter increases progressively with age (TOYOTA et al. 1999b). *Versican* expression is nearly completely suppressed in neoplastic cell lines but can easily be restored to near-normal levels by treatment with DNA-methyltransferase inhibitors. *PAX6*, by contrast, is not known to be expressed in intestinal tissues but is similarly methylated in aging colon (TOYOTA et al. 1999a).

### 3.2.6 Other Genes

Several other genes have been reported to be partially methylated in normal tissues, and it is possible that some of them actually belong to the group methylated in the aging process. *DBCCR1*, just like *N33*, was cloned from an area of frequent loss of heterozygosity on chromosome 9q and was reported to be frequently methylated in bladder cancer and partially methylated in normal bladder (HABUCHI et al. 1998). *E-cadherin* is frequently methylated in normal-appearing liver (KANAI et al. 1997) and stomach (Toyota et al., unpublished) adjacent to cancer, and it is not clear whether this reflects the presence of dysplasia in the area or whether this methylation begins in truly normal cells as a function of age. Other genes hypermethylated in cancer have only been studied in a limited spectrum of normal tissues and were studied using techniques that are not very sensitive. *P15* provides an interesting example of this; this gene is hypermethylated in various hematopoietic malignancies and was found to be non-methylated in normal tissues (HERMAN et al. 1997). Recently, however, using very sensitive techniques, *P15* methylation was detected in a fraction of normal circulating lymphocytes (AGGERHOLM et al. 1999), and it would be interesting to determine whether this is an aging phenomenon as well. Finally, other genes that have been found to be methylated in some normal tissues include *Calcitonin*, mammary-derived-growth-inhibitor gene and others (Herman et al., unpublished).


# 4 The Contribution of Aging to Hypermethylation in Neoplasia


The list of genes affected by age-related methylation is rapidly growing (Table 1). In fact, many genes that had previously been thought to be methylated in cancer exclusively have now been found to belong to this group of age-related genes. Because of this, the process of age-related methylation is a prime suspect in explaining some or all hypermethylation events in cancer. A series of studies have now addressed this issue directly. In a study of the methylation status of six CpG islands in glioblastoma multiforme, three of the six had evidence of being affected by age-related methylation either through partial methylation in normal brain (*HIC1*) or much more frequent methylation in tumors of older patients (*ER* and *N33*; Li et al. 1998). The other three genes were not hypermethylated in normal brain. Furthermore, methylation of one of them, thrombospondin 1, was not associated with either the age of the patient or with methylation of *ER* or *N33* (Li et al. 1999). Thus, in this study, three of six genes hypermethylated in cancer also appeared to be affected by age-related methylation, and the most frequent methylation events in this cancer were those related to aging. A similar study of eight genes hypermethylated in colorectal cancer yielded essentially identical results: four out of these eight genes were affected by age-related methylation in normal colon, and these were the most frequently methylated genes in colorectal cancer (AHUJA et al. 1998).

While the described studies suggested that age-related methylation could account for half of all methylation events in cancer, they may have been biased by gene selection. Several techniques to clone differentially methylated CpG islands have been developed recently and provide a relatively unbiased view of the process. Liang et al. used methylation-sensitive, arbitrarily primed PCR to screen the genome for differentially methylated sequences (LIANG et al. 1998). The majority of

**Table 1.** Genes affected by age-related methylation in human colon

| Gene | Chromosome | Expressed in the colon? | Selected for in cancers? |
|------|-----------|------------------------|--------------------------|
| *EGFR* | 7p12 | Yes | No |
| *ER* | 6q25.1 | Yes | Yes |
| *IGF2* P2–P4 | 11p15.5 | No/low level | Yes |
| *MYOD1* | 11p15.4 | No | Yes |
| *N33* | 8p22 | Yes | Yes |
| *PAX6* | 11p13 | No (?) | Yes |
| *Versican* | 5q12–14 | Yes | Yes |

*EGFR*, epidermal-growth-factor receptor; *ER*, estrogen receptor; *IGF2*, insulin-like-growth-factor-II.

sequences that were hypermethylated in cancer were also methylated to varying degrees in normal tissues. For example, they report that 25 of 45 evaluable bands were differentially methylated in some colon cancers. All but five of these had significant levels of methylation in normal tissues. Similar results were reported for bladder- and prostate-cancer samples. Thus, in this evaluation, 80% of sequences differentially methylated in cancer also had evidence of partial methylation in normal tissues. In a comparable study, Huang et al. used differential methylation hybridization to identify CpG islands differentially methylated in breast cancer (HUANG et al. 1999). Remarkably, 23/237 (9.7%) of all CpG-island fragments examined had some evidence of methylation in normal tissues (some of these may have been derived from the X-chromosome or imprinted genes). Furthermore, they characterized 30 fragments differentially methylated in breast-cancer cell lines and found that 13 (43%) of these were also partially methylated in normal breast. These 13 clones were significantly more frequently methylated in breast cancers than the rest of the fragments studied. While the age of the patients is not specified in that study, analogy to the colon-cancer situation suggests that many of these hypermethylation events will, in fact, be shown to be age-related. Furthermore, because the issue of aging was not addressed specifically (using sensitive methods to compare methylation in older people compared with younger individuals), the study may have underestimated the contribution to the neoplastic process of partial methylation in normal tissues.

Using another method to isolate differentially methylated CpG islands, Toyota et al. cloned 30 DNA fragments from a colon-cancer cell line, some of which corresponded to the 5' region of known genes (TOYOTA et al. 1999b). Twenty-six of these sequences (all of them CpG islands) were methylated in some primary cancers. Of these 26 CpG islands, 19 (73%) were also methylated in normal colon as a function of age (TOYOTA et al. 1999a). Finally, in a simple study of randomly selected, promoter-associated CpG islands, Jair et al. found that seven of 13 genes examined were hypermethylated in some cancer cell lines and primary tumors (Jair et al., in preparation). Of these seven, four were clearly methylated as a function of age in normal colon, including at least one gene that was infrequently methylated in colorectal neoplasms (presumably because of negative selection against methylation of that gene in proliferating cells). All of these studies demonstrate that age-related methylation is a common cause of hypermethylation in cancer, accounting for perhaps as much as 50–80% of all such occurrences.

A critical question raised by these observations is whether all CpG-island methylation in cancer is due to age-related methylation or methylation in subsets of normal cells (such as stem cells) where the neoplastic process begins. Several studies suggest that this is not, in fact, the case. For some genes (such as *P16* and *hMLH1*), methylation cannot be seen in the tissues of older subjects, even using sensitive detection methods (HERMAN et al. 1996, 1998). Furthermore, in colon cancer, methylation of these genes is restricted to a subset of cases that display a profound hypermethylator phenotype (TOYOTA et al. 1999a). Thus, unless the cell of origin is different for colon cancers with this hypermethylator phenotype, one has to assume that these events are truly happening de novo during the development of neoplasia.

In addition, in this same model (colorectal neoplasia), one occasionally detects differences among the methylation status of different regions within the same tumor or between a metastasized tumor and its primary counterpart. Similar differences are observed in leukemias at diagnosis and after relapse from chemotherapy-induced remission (Issa et al. 1997). These examples suggest that the early tumors (and, therefore, the progenitor cell) did not exhibit methylation of the genes under study and that methylation was truly a de novo occurrence in these cases. Interestingly, many (but not all) familial cancer genes belong to this group, where age-related methylation is not observed. This may not be surprising, because age-related methylation (and silencing) of a critical tumor-suppressor gene might be expected to result in a very high incidence of neoplasia and would, therefore, be selected against during evolution.

# 5  Causes of Age-Related Methylation

An understanding of the causes of age-related methylation would shed some light on the pathophysiology of CpG island methylation events in cancer. While there are no definitive answers yet, the growing number of genes affected by this process allows one to make some inferences that are relevant to the causes of this process. It is clear that age-related methylation only affects a subset of human genes, suggesting gene-specific susceptibility to this process. Furthermore, there are considerable tissue-specific differences in the extent of age-related methylation, with indications that the expression state of the involved gene may modulate the process (Ahuja et al. 1998). Indeed, some tissue-specific genes appear to be especially sensitive to age-related methylation in those cells that are not programmed to express the gene. However, ubiquitously expressed genes can also be involved, including some that are expressed at relatively high levels. Because the methylation-free state of CpG islands is thought to result from specific factors that protect the area from de novo methylation (Turker and Bestor 1997), one possibility for reconciliation of these observations is that genes differ in the extent or efficiency of such protection. Thus, genes that are "weakly" protected (a vague term, but little is known about the molecular nature of this protection) may be those that are susceptible to age-related methylation. Furthermore, it is possible that active expression participates in this protection process or may even be required to maintain the methylation-free state of some loci. This, of course, implies that adult cells are capable of de novo methylation, a fact that is increasingly being recognized. Finally, as mentioned above, methylation "spreading" through homologous recombination between methylated sequences (imprinted areas, pseudogenes, etc.) and unmethylated sequences could also play a role in this process.

Differences in extent of methylation among individuals in the same age groups suggest that several factors must also modulate this process. These probably include exposure events (either exogenous carcinogens or endogenously generated reactive oxygen species) and genetic differences in individual susceptibility to age-related methylation. In particular, increased levels of DNA-methyltransferase appear to accelerate this process in vitro (VERTINO et al. 1996), and polymorphisms in the methylation machinery that make it more or less active may be candidates for modulating the process. In addition, it is clear that cell selection also plays an important role in determining the extent of the process. If age-related methylation affects genes that are critical for cell growth or survival (as is likely to be the case if the process is relatively random and depends on local DNA factors), then those cells involved will be strongly selected against over time, which would result in an apparent minimization of the process at these loci. A possible example of this is the epidermal-growth-factor receptor (*EGFR*) oncogene, which is clearly susceptible to some age-related methylation in normal colon and which is highly methylated in hematopoietic malignancies (where the cell of origin does not express this gene). Nevertheless, it is never methylated in colonic tumors, presumably because loss of *EGFR* would result in negative selection (Jair et al., in preparation). Similarly, if age-related methylation affects a gene that negatively regulates growth or promotes apoptosis, then the cells where this process begins may have a selective advantage over neighboring cells, resulting in an amplification of the methylation phenomenon.

Finally, it is unclear whether there are interactions between methylation and other physiologic changes that accompany aging (GREIDER 1990; SMITH and PEREIRASMITH 1996). In particular, it is not known whether there are interactions between telomere shortening or telomerase activation and CpG-island methylation. Similarly, aging is associated with changes in gene expression, some of which may affect methylation. Indeed, a decrease in gene expression may set up a CpG island for hypermethylation, and there may be genes important for protection against de novo methylation whose expression declines in aging cells. Of course, as discussed below, hypermethylation itself may be the cause of some of these age-related declines in gene expression. Overall, it appears safe to assume that age-related methylation is caused by factors that are inherent to DNA structure, combined with modulating factors, such as exposure and predisposition.

# 6 Functional Implications of Age-Related Methylation

The genes that are affected by age-related methylation fall into two general categories (Table 1). One group of genes consists of those that are expressed in a tissue-specific manner and that become methylated in non-expressing tissues. In these cases, methylation simply marks the transcriptional silent state and is unlikely to change the physiology of the involved tissues. The second group of genes, however, consists of widely expressed genes for which age-related methylation may actually result in transcriptional silencing. The methylation–transcription link is fairly well established for this group, which includes *ER*, *N33* and *Versican*, and the loss of expression of these genes in methylated cells is not due to loss of transcription factors, as some of them are relatively easily reactivated by treatment with methylation inhibitors.

It is not clear whether age-related methylation, which is invariably reflected by partial methylation, affects all cells in a given tissue (with relatively sparse methylation) or only a small subset of cells, where methylation could be bi-allelic and heavy. Current evidence from cell-culture experiments and genomic sequencing favors the latter hypothesis. For example, Vertino et al. found that *ER* methylation in normal fibroblasts reflects the average of multiple clones, some of which have high levels of methylation and others of which have nearly undetectable methylation (VERTINO et al. 1994). The implication of these findings is that aging tissues have a distinct subset of cells with altered physiology (through reduced gene expression) when compared with normal cells. Furthermore, because these cells with hypermethylated genes are particularly prominent in early stages of neoplasia, such as adenomatous polyps of the colon (ISSA et al. 1994), it appears that they may be predisposed to neoplastic transformation. This predisposition may, in fact, stem from the age-related methylation process, particularly if the genes affected influence the growth, differentiation or apoptotic status of normal cells. Age-related methylation, therefore, has the potential to behave as a mutator process, resulting in the simultaneous silencing of multiple genes in aging tissues. Those cells affected may well represent the "field defect" observed in patients with neoplasia, whereby the risk of additional neoplasms is very high following an initial diagnosis of cancer (LIPKIN 1988). In fact, the age-related emergence and expansion of this field defect could account, in part, for the marked and exponential increase in cancer incidence with age (Fig. 2).

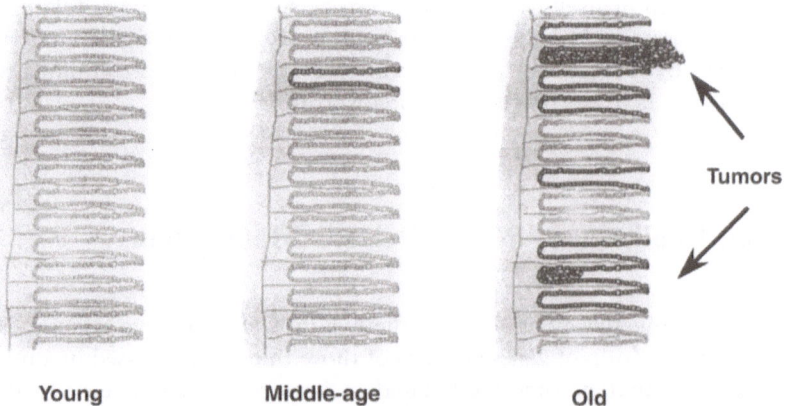

Young                Middle-age              Old

**Fig. 2.** A hypothetical model of the role of age related methylation in colorectal cancer pathogenesis. In young individuals, the colorectal epithelium is unmethylated and intact (grey cells, *left panel*). With age, some crypts acquire promoter methylation and silencing at selected loci (black cells, *middle panel*), which results in a disturbed physiology that includes increased proliferation. The crypts where methylation is found are those where tumors eventually form through the acquisition of further molecular defects (*right*)

The model whereby age-related methylation marks (and partly leads to) cancer predisposition implies two testable hypotheses: (1) quantitative measurement of age-related methylation may be useful in risk assessment for neoplastic disease, and (2) modulation of this process may result in changes in the risk of developing pre-neoplastic and neoplastic lesions. Testing of the first hypothesis requires relatively large studies, and such investigations are currently underway. However, an important factor here is whether age-related methylation is a rate-limiting step in neoplastic transformation (in which case it should correlate well with tumor incidence) or whether it is so prevalent that the rate-limiting steps are actually subsequent events (mutations in gatekeeper genes, etc.). The second hypothesis has already had a test of sorts. Laird et al. have reported that lowering DNA methylation reduces tumor incidence in the Min mouse model of colorectal neoplasia (LAIRD et al. 1995), and it is likely that the mechanism of this preventive effect involves, in part, reduced age-related methylation in this system. Therefore, in addition to predictive testing, one of the potentially critical clinical implications of the process of age-related methylation is that it could be an excellent target for preventive approaches in neoplasia.

Finally, as discussed above, age-related methylation is not limited to particular tissues or cells and, indeed, is not an exclusive property of the neoplastic process. An intriguing possibility, then, is that aberrant methylation could contribute to other age-related diseases as well. Several of these are accompanied by dysregulated growth (such as reactive smooth-muscle cells in atherosclerosis plaques) or decreased gene function (such as the progressive age-related insulin resistance observed in some people, which may be a harbinger of adult-onset diabetes mellitus), and these could, theoretically, be due to age-related methylation. Further studies should address these issues directly.

# 7 Conclusions

It has become clear over the past few years that a significant fraction of human genes is affected by a process of age-related promoter methylation that has the potential to silence gene expression, alter the physiology of aging cells and possibly predispose them to neoplastic transformation. One ominous conclusion of all these observations is that age-related methylation represents a kind of molecular clock attached to a time bomb, with disastrous consequences for the integrity of aged tissues. A redeeming feature of this gloomy prediction is that methylation is an enzymatic process that can be measured, slowed and possibly reversed using various strategies. Indeed, the identification of methylation as a prominent feature of aging and associated diseases opens new areas of research in risk assessment and, perhaps most importantly, in disease prevention.

*Acknowledgements.* Work in the author's laboratory is supported by grants from the National Institute of Health (USA), the American Cancer Society and the Kimmel Foundation.

# References

Aggerholm A, Guldberg P, Hokland M, Hokland P (1999) Extensive intra- and interindividual heterogeneity of p15INK4B methylation in acute myeloid leukemia. Cancer Res 59:436–441

Ahuja N, Li Q, Mohan AL, Baylin SB, Issa JPJ (1998) Aging and DNA methylation in colorectal mucosa and cancer. Cancer Res 58:5489–5494

Barlow DP (1995) Gametic imprinting in mammals. Science 270:1610–1613

Baylin SB, Herman JG, Graff JR, Vertino PM, Issa JPJ (1998) Alterations in DNA methylation – a fundamental aspect of neoplasia. Adv Cancer Res 72:141–196

Bestor TH, Tycko B (1996) Creation of genomic methylation patterns. Nat Genet 12:363–367

Bird AP (1986) CpG-rich islands and the function of DNA methylation. Nature 321:209–213

Chen RZ, Pettersson U, Beard C, Jackson-Grusby L, Jaenisch R (1998) DNA hypomethylation leads to elevated mutation rates. Nature 395:89–93

Choi EK, Uyeno S, Nishida N, Okumoto T, Fujimura S, Aoki Y, Nata M, Sagisaka K, Fukuda Y, Nakao K, Yoshimoto T, Kim YS, Ono T (1996) Alterations of *c-fos* gene methylation in the processes of aging and tumorigenesis in human liver. Mutat Res 354:123–128

Davis RL, Weintraub H, Lassar AB (1987) Expression of a single transfected cDNA converts fibroblasts to myoblasts. Cell 51:987–1000

Drinkwater RD, Blake TJ, Morley AA, Turner DR (1989) Human lymphocytes aged in vivo have reduced levels of methylation in transcriptionally active and inactive DNA. Mutat Res 219:29–37

Ferguson AT, Lapidus RG, Baylin SB, Davidson NE (1995) Demethylation of the estrogen receptor gene in estrogen receptor-negative breast cancer cells can reactivate estrogen receptor gene expression. Cancer Res 55:2279–2283

Flavell RB (1994) Inactivation of gene expression in plants as a consequence of specific sequence duplication. Proc Natl Acad Sci USA 91:3490–3496

Fujii H, Biel MA, Zhou W, Weitzman SA, Baylin SB, Gabrielson E (1998) Methylation of the HIC-1 candidate tumor suppressor gene in human breast cancer. Oncogene 16:2159–2164

Gama-Sosa MA, Midgett RM, Slagel VA, Githens S, Kuo KC, Gehrke CW, Ehrlich M (1983) Tissue-specific differences in DNA methylation in various mammals. Biochim Biophys Acta 740: 212–219

Ghazi H, Gonzales FA, Jones PA (1992) Methylation of CpG-island-containing genes in human sperm, fetal and adult tissues. Gene 114:203–210

Giannoukakis N, Deal C, Paquette J, Goodyer CG, Polychronakos C (1993) Parental genomic imprinting of the human *igf2* gene. Nat Genet 4:98–101

Goodman JI, Counts JL (1993) Hypomethylation of DNA: a possible nongenotoxic mechanism underlying the role of cell proliferation in carcinogenesis. Environ Health Perspect 101, Suppl 5:169–172

Goto T, Monk M (1998) Regulation of X-chromosome inactivation in development in mice and humans. Microbiol Mol Biol Rev 62:362–378

Greider CW (1990) Telomeres, telomerase and senescence. Bioessays 12:363–369

Habuchi T, Luscombe M, Elder PA, Knowles MA (1998) Structure and methylation-based silencing of a gene (DBCCR1) within a candidate bladder cancer tumor suppressor region at 9q32–q33. Genomics 48:277–288

Herman JG, Graff JR, Myohanen S, Nelkin BD, Baylin SB (1996) Methylation-specific PCR: a novel PCR assay for methylation status of CpG islands. Proc Natl Acad Sci USA 93:9821–9826

Herman JG, Civin CI, Issa JPJ, Collector MI, Sharkis SJ, Baylin SB (1997) Distinct patterns of p15INK4B and p16INK4A characterize the major types of hematologic malignancies. Cancer Res 57:837–841

Herman JG, Umar A, Polyak K, Graff JR, Ahuja N, Issa JPJ, Markowitz S, Willson JKV, Hamilton SR, Kinzler KW, Kane MF, Kolodner RD, Vogelstein B, Kunkel TA, Baylin SB (1998) Incidence and functional consequences of hMLH1 promoter hypermethylation in colorectal carcinoma. Proc Natl Acad Sci USA 95:6870–6875

Hoal-van Helden EG, van Helden PD (1989) Age-related methylation changes in DNA may reflect the proliferative potential of organs. Mutat Res 219:263–266

Holliday R (1985) The significance of DNA methylation in cellular aging. Basic Life Sci 35:269 283

Holliday R (1986) Strong effects of 5-azacytidine on the in vitro lifespan of human diploid fibroblasts. Exp Cell Res 166:543–552

Holliday R (1987) The inheritance of epigenetic defects. Science 238:163–170

Huang THM, Perry MR, Laux DE (1999) Methylation profiling of CpG islands in human breast cancer cells. Hum Mol Genet 8:459–470

Issa JP, Ottaviano YL, Celano P, Hamilton SR, Davidson NE, Baylin SB (1994) Methylation of the oestrogen receptor CpG island links ageing and neoplasia in human colon. Nat Genet 7:536–540

Issa JPJ, Baylin SB, Belinsky SA (1996a) Methylation of the estrogen receptor CpG island in lung tumors is related to the specific type of carcinogen exposure. Cancer Res 56:3655–3658

Issa JPJ, Vertino PM, Boehm CD, Newsham IF, Baylin SB (1996b) Switch from mono-allelic to bi-allelic human IGF2 promoter methylation during aging and carcinogenesis. Proc Natl Acad Sci USA 93:11757–11762

Issa JPJ, Zehnbauer BA, Civin CI, Collector MI, Sharkis SJ, Davidson NE, Kaufmann SH, Baylin SB (1996c) The estrogen receptor CpG island is methylated in most hematopoietic neoplasms. Cancer Res 56:973–977

Issa JPJ, Zehnbauer BA, Kaufmann SH, Biel MA, Baylin SB (1997) HIC1 hypermethylation is a late event in hematopoietic neoplasms. Cancer Res 57:1678–1681

Jones PA, Laird PW (1999) Cancer epigenetics comes of age. Nat Genet 21:163–167

Kanai Y, Ushijima S, Hui AM, Ochiai A, Tsuda H, Sakamoto M, Hirohashi S (1997) The E-cadherin gene is silenced by CpG methylation in human hepatocellular carcinomas. Int J Cancer 71:355–359

Kass SU, Pruss D, Wolffe AP (1997) How does DNA methylation repress transcription? Trends Genet 13:444–449

Laird PW, Jackson-Grusby L, Fazeli A, Dickinson SL, Jung WE, Li E, Weinberg RA, Jaenisch R (1995) Suppression of intestinal neoplasia by DNA hypomethylation. Cell 81:197–205

Lasalle JM, Lalande M (1996) Homologous association of oppositely imprinted chromosomal domains. Science 272:725–728

Li Q, Jedlicka A, Ahuja N, Gibbons MC, Baylin SB, Burger PC, Issa JPJ (1998) Concordant methylation of the ER and N33 genes in glioblastoma multiforme. Oncogene 16:3197–3202

Li Q, Ahuja N, Burger PC, Issa JPJ (1999) Methylation and silencing of the thrombospondin-1-promoter in human cancer. Oncogene 18:3284–3289

Liang G, Salem CE, Yu MC, Nguyen HD, Gonzales FA, Nguyen TT, Nichols PW, Jones PA (1998) DNA methylation differences associated with tumor tissues identified by genome scanning analysis. Genomics 53:260–268

Lipkin M (1988) Biomarkers of increased susceptibility to gastrointestinal cancer: new application to studies of cancer prevention in human subjects. Cancer Res 48:235–245

Losordo DW, Kearney M, Kim EA, Jekanowski J, Isner JM (1994) Variable expression of the estrogen receptor in normal and atherosclerotic coronary arteries of premenopausal women. Circulation 89:1501–1510

MacGrogan D, Levy A, Bova GS, Isaacs WA, Bookstein R (1996) Structure and methylation-associated silencing of a gene within a homozygously deleted region of human chromosome band 8p22. Genomics 35:55–65

Makos M, Nelkin BD, Reiter RE, Gnarra JR, Brooks J, Isaacs W, Linehan M, Baylin SB (1993) Regional DNA hypermethylation at D17S5 precedes 17p structural changes in the progression of renal tumors. Cancer Res 53:2719–2722

Maloisel L, Rossignol JL (1998) Suppression of crossing-over by DNA methylation in *Ascobolus*. Genes Dev 12:1381–1389

Mays-Hoopes L, Chao W, Butcher HC, Huang RC (1986) Decreased methylation of the major mouse long interspersed repeated DNA during aging and in myeloma cells. Dev Genet 7:65–73

Mendelsohn ME, Karas RH (1994) Estrogen and the blood vessel wall. Curr Opin Cardiol 9:619–626

Miyamura Y, Tawa R, Koizumi A, Uehara Y, Kurishita A, Sakurai H, Kamiyama S, Ono T (1993) Effects of energy restriction on age-associated changes of DNA methylation in mouse liver. Mutat Res 295:63–69

Morton RA, Jr Watkins JJ, Bova GS, Wales MM, Baylin SB, Isaacs WB (1996) Hypermethylation of chromosome 17P locus D17S5 in human prostate tissue. J Urol 156(Pt 1):512–516

Nathan L, Chaudhuri G (1997) Estrogens and atherosclerosis. Annu Rev Pharmacol Toxicol 37:477–515

Ono T, Tawa R, Shinya K, Hirose S, Okada S (1986) Methylation of the *c-myc* gene changes during aging process of mice. Biochem Biophys Res Commun 139:1299–1304

Ono T, Takahashi N, Okada S (1989) Age-associated changes in DNA methylation and mRNA level of the *c-myc* gene in spleen and liver of mice. Mutat Res 219:39–50

Ono T, Uehara Y, Kurishita A, Tawa R, Sakurai H (1993) Biological significance of DNA methylation in the ageing process. Age Ageing 22:S34–43

Ottaviano YL, Issa JP, Parl FF, Smith HS, Baylin SB, Davidson NE (1994) Methylation of the estrogen receptor gene CpG island marks loss of estrogen receptor expression in human breast cancer cells. Cancer Res 54:2552–2555

Post WS, Clermont-Goldschmidt PJ, Heldman AW, Sussman MS, Ouyangan P, Milliken EE, Issa JPJ (1999) Methylation of the estrogen-receptor gene in cardiovascular tissue is related to aging and atherosclerosis. Cardiovascular Res 43:985–991

Potter JD, Slattery ML, Bostick RM, Gapstur SM (1993) Colon cancer: a review of the epidemiology. Epidemiol Rev 15:499–545

Razin A, Shemer R (1995) DNA methylation in early development. Hum Mol Genet 4, Spec No:1751–5

Rideout WM, Eversole-Cire P, Spruck CH, Hustad CM, Coetzee GA, Gonzales FA, Jones PA (1994) Progressive increases in the methylation status and heterochromatinization of the MyoD CpG island during oncogenic transformation. Mol Cell Biol 14:6143–6152

Riggs AD, Xiong Z, Wang L, LeBon JM (1998) Methylation dynamics, epigenetic fidelity and X chromosome structure. Novartis Found Symp 214:214–225

Rohde M, Warthoe P, Gjetting T, Lukas J, Bartek J, Strauss M (1996) The retinoblastoma protein modulates expression of genes coding for diverse classes of proteins including components of the extracellular matrix. Oncogene 12:2393–2401

Romanov GA, Vanyushin BF (1981) Methylation of reiterated sequences in mammalian DNAs. Effects of the tissue type, age, malignancy and hormonal induction. Biochim Biophys Acta 653:204–218

Roncucci L, Ponz de Leon M, Scalmati A, Malagoli G, Pratissoli S, Perini M, Chahin NJ (1988) The influence of age on colonic epithelial cell proliferation. Cancer 62:2373–2377

Shemer R, Kafri T, O'Connell A, Eisenberg S, Breslow JL, Razin A (1991) Methylation changes in the apolipoprotein AI gene during embryonic development of the mouse. Proc Natl Acad Sci USA 88:11300–11304

Smith JR, Pereirasmith OM (1996) Replicative senescence – implications for in vivo aging and tumor suppression. Science 273:63–67

Sussenbach JS, Rodenburg RJ, Scheper W, Holthuizen P (1993) Transcriptional and post-transcriptional regulation of the human IGF-II gene expression. Adv Exp Med Biol 343:63–71

Tornaletti S, Pfeifer GP (1995) Complete and tissue-independent methylation of CpG sites in the p53 gene: implications for mutations in human cancers. Oncogene 10:1493–1499

Toyota M, Ahuja N, Ohe-Toyota M, Herman JG, Baylin SB, Issa JPJ (1999a) CpG-island-methylator phenotype in colorectal cancer. Proc Natl Acad Sci USA 96:8681–8686

Toyota M, Ho C, Ahuja N, Jair K-W, Ohe-Toyota M, Baylin SB, Issa JPJ (1999b) Identification of differentially methylated sequences in colorectal cancer by methylated CpG-island amplification. Cancer Res 59:2307–2312

Turker MS, Bestor TH (1997) Formation of methylation patterns in the mammalian genome. Mutat Res 386:119–130

Vertino PM, Issa JP, Pereira-Smith OM, Baylin SB (1994) Stabilization of DNA methyltransferase levels and CpG island hypermethylation precede SV40-induced immortalization of human fibroblasts. Cell Growth Differ 5:1395–1402

Vertino PM, Yen RW, Gao J, Baylin SB (1996) De novo methylation of CpG island sequences in human fibroblasts overexpressing DNA (cytosine-5-)-methyltransferase. Mol Cell Biol 16:4555–4565

Vu TH, Hoffman AR (1994) Promoter-specific imprinting of the human insulin-like growth-factor-II gene. Nature 371:714–717

Wales MM, Biel MA, el Deiry W, Nelkin BD, Issa JP, Cavenee WK, Kuerbitz SJ, Baylin SB (1995) p53 Activates expression of HIC-1, a new candidate tumour suppressor gene on 17p13.3. Nat Med 1: 570–577

Walsh CP, Bestor TH (1999) Cytosine methylation and mammalian development. Genes Dev 13:26–34

Wilson VL, Jones PA (1983a) DNA methylation decreases in aging but not in immortal cells. Science 220:1055–1057

Wilson VL, Jones PA (1983b) Inhibition of DNA methylation by chemical carcinogens in vitro. Cell 32:239–246

Wilson VL, Smith RA, Ma S, Cutler RG (1987) Genomic 5-methyldeoxycytidine decreases with age. J Biol Chem 262:9948–9951

# Mouse Models in DNA-Methylation Research

P.W. Laird

# 1 Introduction

DNA-methylation abnormalities in human cancer cells have been described in numerous reports for close to two decades, providing an extensive and powerful argument for an involvement of DNA methylation in oncogenesis (Jones and Laird 1999). However, these studies do not address the nature of the relationship. Does cancer affect DNA methylation or does DNA methylation influence cancer – or both? Experimental models that allow the manipulation of DNA-methylation patterns can provide insight into the causal direction of the relationship. They are

Departments of Surgery and of Biochemistry and Molecular Biology, University of Southern California Keck School of Medicine, Norris Comprehensive Cancer Center, Room 6418, 1441 Eastlake Ave., Los Angeles, CA 90089-9176, USA
E-mail: plaird@hsc.usc.edu

also instrumental in understanding the mechanistic underpinnings of the relationship between DNA methylation and cancer.

Various techniques have been used to experimentally modify methylation patterns in cells and live organisms. The most widely employed approach has been to reduce DNA-methylation levels globally by limiting the methyl-group supply or by manipulating the active levels of the DNA-methyltransferase enzymes that catalyze the transfer of methyl groups to unmethylated cytosine substrates. DNA-methyltransferase-activity levels can be modulated by drug inhibition, by oligonucleotide or antisense technology and by transgenic or gene targeting techniques. This chapter will examine these various experimental approaches. The main emphasis will be on the use of gene-knockout technology in mice, which has contributed most extensively to our understanding of the role of DNA methylation in mammals.

# 2 Mammalian DNA Methyltransferases and Demethylases

Over the last decade, only a single mammalian DNA (cytosine-5)-methyltransferase had been characterized and cloned. Therefore, the older literature refers to the gene encoding this major enzyme simply as the mammalian DNA methyltransferase gene, the MTase gene or, later, the *Dnmt* gene. Once additional candidate DNA (cytosine-5)-methyltransferases had been identified, the *Dnmt* gene was re-designated as *Dnmt1* in mice and *DNMT1* in humans. DNA (cytosine-5)-methyltransferases are recognizable by a set of conserved motifs, which led to the cloning of several new candidate DNA-methyltransferase genes, *Dnmt2*, *Dnmt3a* and *Dnmt3b*. Although it contains the set of conserved motifs, the *Dnmt2* gene has, so far, not been shown to encode a catalytically active DNA (cytosine-5)-methyltransferase. Recently, the first mammalian DNA (5-methylcytosine)-demethylase gene was cloned and sequenced.

## 2.1 *Dnmt1*

The *Dnmt1* gene encodes the predominant DNA (cytosine-5)-methyltransferase present in mammalian cells. The enzyme is unusual in that it contains a large N-terminal extension not found in other DNA (cytosine-5)-methyltransferases. This N-terminal domain contains several interesting functional elements (Fig. 1), including a nuclear localization signal, replication-foci-targeting sequences (LEON-HARDT et al. 1992; LIU et al. 1998), a zinc-binding domain (BESTOR 1992), a major serine-phosphorylation site (GLICKMAN et al. 1997) and a proliferating-cell-nuclear-antigen-interacting motif (CHUANG et al. 1997). The first mouse and human complementary DNA (cDNA) clones of the *Dnmt1* gene were eventually found to be incomplete, and additional upstream exons were subsequently described for both

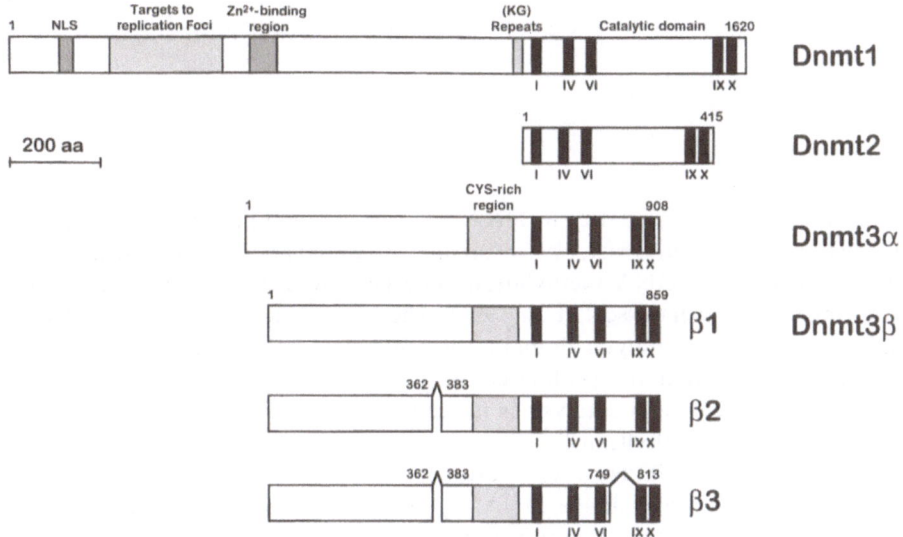

**Fig. 1.** Schematic illustrating the structure of the murine DNA (cytosine-5)-methyltransferase proteins, drawn approximately to scale. Significant protein motifs are indicated by *shaded boxes*. The conserved motifs in the catalytic domain are indicated by *roman numerals*. NLS indicates the nuclear localization signal. Amino acid positions are indicated by the *numbering* above the *bars*. This figure is adapted from OKANO et al. (1998a)

the human and mouse genes (TUCKER et al. 1996b; YODER et al. 1996; GLICKMAN et al. 1997). The first exon is alternatively spliced in post-mitotic germ cells (MERTINEIT et al. 1998). An upstream exon, exon 1o, is utilized in growing oocytes, while an exon further downstream, exon 1p, which interferes with translation, is used in pachytene spermatocytes. In between these two sex-specific exons lies the somatic exon 1s, which is used in other cell types.

A particularly remarkable feature of the Dnmt1 protein is that its de novo methyltransferase activity is much lower than its maintenance DNA-methyltransferase activity. These two distinct types of DNA methyltransfer reactions can be distinguished based on the methylation status of the target sequence. Duplex DNA that has undergone recent replication consists of a methylated parent strand and an unmethylated daughter DNA strand. Methylation of target sequences in the daughter strand of such hemi-methylated DNA is referred to as maintenance DNA methylation, while methylation of previously unmethylated target substrates is referred to as de novo methylation. All known DNA (cytosine-5)-methyltransferases are capable of performing both reactions, usually with equal efficiencies. The de novo methyltransferase activity of the Dnmt1 enzyme is substantially suppressed compared with its maintenance DNA-methyltransferase activity. This makes the Dnmt1 enzyme well suited for the faithful transmission of DNA-methylation patterns during cell division, without resulting in widespread de novo methylation of unmethylated sites in the genome. While there is evidence for the de novo methylating capability of the Dnmt1 protein in vivo (YODER et al. 1997), its major task appears to be the post-replicative restoration of hemi-methylated sites to full methylation.

## 2.2 *Dnmt2*

A gene with sequence similarity to DNA (cytosine-5)-methyltransferase genes was identified in the fission yeast *Schizosaccharomyces pombe*, an organism lacking detectable cytosine-5 DNA methylation. The gene was named *pmt1* (for *pombe* methyltransferase; WILKINSON et al. 1995). The *pmt1* gene contains the conserved motifs found in the catalytic domains of DNA (cytosine-5)-methyltransferases, with the exception that the proline codon in the essential proline–cysteine (PC) motif involved in covalent catalysis is instead a serine codon. Indeed, attempts to demonstrate DNA-methyltransferase activity for this enzyme were unsuccessful (WILKINSON et al. 1995). However, restoration of the PC motif in this enzyme resulted in a catalytically active DNA (cytosine-5)-methyltransferase (PINARBASI et al. 1996). Subsequently, several groups reported a mammalian gene with close sequence similarity to the *pmt1* gene. This new putative DNA (cytosine-5)-methyltransferase gene was designated Dnmt2 in mice and *DNMT2* in humans (OKANO et al. 1998b; VAN DEN WYNGAERT et al. 1998; YODER and BESTOR 1998). Interestingly, both the human and mouse versions contain the correct PC motif, in addition to the other conserved motifs (Fig. 1). Nevertheless, neither *Escherichia coli*-expressed human DNMT2 protein, nor baculovirus-expressed mouse *Dnmt2* protein show detectable transmethylase activity (OKANO et al. 1998b; YODER and BESTOR 1998). The function of the Dnmt2 enzyme remains an enigma, but it has been proposed to be a DNA (cytosine-5)-methyltransferase with high de novo activity and unusual substrate or co-factor requirements (OKANO et al. 1998b; YODER and BESTOR 1998). However, its expression pattern is not restricted to cell types with active de novo methylation (OKANO et al. 1998b; YODER and BESTOR 1998). Importantly, embryonic stem (ES) cells with a homozygous targeted deletion of the PC motif in the catalytic domain of the *Dnmt2* gene do not show any change in their DNA-methylation levels or in their ability to perform de novo methylation (OKANO et al. 1998b). It should be noted, however, that these cells still have functional *Dnmt1* methyltransferase, which could well mask a methylation defect associated with *Dnmt2* deficiency. Resolution of this issue may require the generation of ES cells with a double knockout of both *Dnmt1* and *Dnmt2*.

## 2.3 *Dnmt3a* and *Dnmt3b*

The family of mammalian DNA (cytosine-5)-methyltransferase genes was expanded with the identification of two closely related genes, *Dnmt3a* and *Dnmt3b* (OKANO et al. 1998a). Both genes are conserved in mouse and human and show little sequence similarity to either *Dnmt1* or *Dnmt2*. The encoded proteins Dnmt3α and Dnmt3β have been demonstrated to be catalytically active, with approximately equal ratios of de novo to maintenance DNA methyltransferase activity (Fig. 1). This feature distinguishes these enzymes from Dnmt1, which has suppressed de novo methyltransferase activity. The *Dnmt3a* and *Dnmt3b* genes are abundantly expressed in undifferentiated ES cells but less strongly expressed in differentiated cells and in adult tissues (ROBERTSON et al. 1999). Their expression pattern is consistent with a role in the wave of post-implantation de novo methylation that occurs during mouse embryogenesis. Three alternative splice forms of *Dnmt3β* exist (Fig. 1), at least two of which have catalytic activity (OKANO et al. 1998a). In addition, two different *Dnmt3a* transcript sizes have been detected.

## 2.4 Demethylases

Passive loss of DNA methylation occurs when a fully methylated CpG dinucleotide goes through two rounds of DNA replication without intermittent maintenance DNA methylation. In contrast, active demethylation refers to the enzymatic, replication-independent replacement of 5-methylcytosine by unmethylated cytosine. Various molecular mechanisms have been proposed, such as DNA-glycosylase activity, dinucleotide exchange and removal of only the methyl group (WEISS et al. 1996; JOST et al. 1997; SWISHER et al. 1998; BHATTACHARYA et al. 1999; CEDAR and VERDINE 1999). The only activity for which a gene has been unambiguously cloned and sequenced is a direct demethylase capable of abstracting a methyl group from 5-methylcytosine incorporated into DNA, to yield unmethylated cytosine and methanol as products (BHATTACHARYA et al. 1999; CEDAR and VERDINE 1999).

# 3  Experimental Manipulation of DNA-Methylation Levels

Normal endogenous mammalian cytosine-5 DNA methylation arises in a post-replicative process that involves the genomic DNA substrate, an *S*-adenosylmethionine (SAM) methyl group donor (Ado-Met) and a DNA (cytosine-5)-methyltransferase enzyme. Any one of these three components can be a target for experimental manipulation of either the global level of DNA methylation or of the pattern of 5-methylcytosine distribution.

## 3.1  Targeting the DNA Substrate

DNA-methylation levels can be manipulated directly without affecting the endogenous methyltransferases or the methyl-group supply. One straightforward way is to introduce an exogenous DNA (cytosine-5)-methyltransferase gene. However, it has been notoriously difficult to achieve expression of exogenous DNA (cytosine-5)-methyltransferase genes, both in cell culture and in transgenic mice. Expression of prokaryotic DNA (cytosine-5)-methyltransferases, such as HhaI, is transforming and lethal to NIH/3T3 cells (Wu et al. 1996). Even overexpression of exogenously introduced *Dnmt1* genes has been problematic. Modest (twofold) overexpression of *Dnmt1* induces transformation of NIH/3T3 cells (Wu et al. 1993). However, this study was performed with an incomplete cDNA of the *Dnmt1* gene (Tucker et al. 1996b; Yoder et al. 1996). Expression of such a truncated version of the murine gene is tolerated only at very low levels in mouse ES cells (Tucker et al. 1996b), whereas a mini-gene including upstream exons can be expressed at higher levels (Tucker et al. 1996b). One caveat to the conclusion that the truncation is responsible for the apparent toxicity is that the successful expression of the mini-gene was driven from the endogenous *Dnmt1* promoter. Therefore, an alternative interpretation is that excessive expression or non-cell-cycle-regulated expression, rather than N-terminal truncation of Dnmt1, is deleterious to the cell. Substantial overexpression of the human *DNMT1* gene was achieved in SV40 immortalized human fetal lung fibroblasts, leading to significant de novo methylation of endogenous CpG-island sequences (Vertino et al. 1996). The human *DNMT1* cDNA used in this study was thought to be full length at the time but later turned out to lack 5' sequences (Yoder et al. 1996). It is not clear why these immortalized human cells tolerated high levels of expression of the truncated human *DNMT1* gene while mouse ES cells did not tolerate even modest expression levels of a similarly truncated murine *Dnmt1* construct.

A drawback to the strategy above is that little control can be exerted on which sequences are targeted for de novo methylation. A first and ingenious step towards a more directed methylation of specific target sequences was taken with the development of hybrid proteins consisting of a prokaryotic DNA (cytosine-5)-methyltransferase moiety and zinc fingers with novel DNA-recognition capabilities (Xu and Bestor 1997). The fusion of the zinc finger to the DNA methyltransferase reduced the affinity of the enzyme for generic CpG-containing DNA, but this was offset by an increase in the local enzyme concentration in the vicinity of the zinc-finger-binding site. The net result is an enzyme with an enhanced sequence specificity for the methyltransfer reaction. Further development could lead to DNA (cytosine-5)-methyltransferase enzymes with unique target-substrate sequences.

An unusual method to directly manipulate endogenous 5-methylcytosine pools is by uptake and incorporation of exogenous 5-methyldeoxycytidine. This was first demonstrated by microinjection into sea-urchin embryos (Chen et al. 1993) and has now been extended to include spontaneous uptake by tissue-culture cells (Holliday and Ho 1998). Aside from the interesting question of whether this phenomenon occurs normally in the salvage of 5-methyl-deoxycytosine monophosphate released as a consequence of DNA repair, it does appear that some 5-methyldeoxycytidine from exogenous sources can be incorporated into genomic DNA, much of it as thymine. Most of the 5-methylcytosine that is incorporated as such from exogenous 5-methyldeoxycytidine sources would not be within a CpG-sequence context and, therefore, probably would not be stably maintained.

It has also been reported that exposure of cells or tissues to anticancer agents can lead to an increase of global 5-methylcytosine levels (Nyce 1997). The mechanism responsible for this effect has not been resolved, but it is thought that aberrant DNA structures induced by the agents may trigger de novo methylation.

## 3.2 Targeting the Methyl-Group Supply

Manipulation of the methyl-group supply has been exploited extensively both in cell-culture systems and in vivo. The methyl donor SAM is converted to S-adenosylhomocysteine (SAH) during the methyl-transfer reaction. SAH is subsequently converted to homocysteine, which is then re-methylated to yield methionine in a folate- and cobalamin-dependent reaction before being converted back to SAM. This cycle can be interrupted by depletion of the methionine backbone or by blocking the folate-dependent regeneration of methionine from homocysteine. Methionine depletion is severely growth inhibitory to cells and animals and is, therefore, usually combined with replenishment of the unmethylated methionine backbone in the form of L-homocystine (the disulfide version of the more toxic L-homocysteine). This regimen is often combined with folate deprivation, which inhibits the regeneration of methionine from homocysteine. This exacerbates the methyl-group deficiency without further depleting the combined methionine/homocysteine-backbone levels. The methionine antimetabolite ethionine has similar SAM-depleting effects. Glutathione-depleting drugs can also lead to SAM deficiency since the re-synthesis of glutathione occurs at the expense of methionine (LERTRATANANGKOON et al. 1996; LERTRATANANGKOON et al. 1997).

Dietary methyl-group deficiency has been widely used in rat and mouse models of liver carcinogenesis (POIRIER 1994; POGRIBNY et al. 1995, 1997; COUNTS et al. 1997). The underlying oncogenic mechanism of methyl-group depletion has not been fully resolved. Possibilities that have been proposed include an increased rate of deletion or mutation (POGRIBNY et al. 1995; CHEN et al. 1998), an increased expression of proto-oncogenes (COUNTS and GOODMAN 1995) and DNA-methyl-transferase-induced cytosine deamination (SHEN et al. 1992; LAIRD and JAENISCH 1994). However, it should be emphasized that methyl-group depletion affects many other forms of biological methylation. The methylation of cytosine in DNA represents only a small fraction of the biological transmethylation that occurs within the cell.

The in vivo manipulation of SAM pools can be achieved by genetic means, as can the dietary depletion of the methyl-group supply outlined above. Mice deficient in adenosine deaminase develop elevated levels of adenosine, which leads to product inhibition of SAH hydrolase (MIGCHIELSEN et al. 1995). The associated increase of SAH or decrease in the SAM/SAH ratio is thought to inhibit SAM-dependent transmethylation reactions. A mouse mutant with a deletion of the SAH hydrolase gene has also been described (MILLER et al. 1994). Homozygosity of either of these mutations leads to embryonic lethality, which has limited their use in DNA-methylation research.

## 3.3 Targeting the DNA (Cytosine-5)-Methyltransferases

The mammalian DNA (cytosine-5)-methyltransferase enzymes offer excellent targets for the manipulation of DNA-methylation levels in cell cultures and in vivo (JACKSON-GRUSBY and JAENISCH 1996). These enzymes have been the targets of manipulation at the gene level using knockout technology, at the transcript level using antisense technology, and at the protein level using mechanism-based inhibitors.

### 3.3.1 Gene Targeting

Gene-knockout technology relies on efficient homologous recombination in mouse ES cells to introduce mutations into specific gene loci for study in cell culture and in vivo. Several different knockout alleles of the mouse *Dnmt1* gene have been generated (Table 1). The most widely studied mutation is a $Neo^R$ insertion near the 5' end of the gene, which deletes N-terminal sequences (LI et al. 1992). The full designation of this allele is $Dnmt1^{tm1Jae}$, but it is widely referred to as the $Dnmt1^n$ allele for the NaeI deletion that is created (or for "N-terminal disruption"; LEI et al. 1996). $Dnmt1^{n/n}$ mice die at around day 10.5 of embryonic development. $Dnmt1^{n/n}$ embryos show severe DNA hypomethylation and are stunted and developmentally delayed.

The direct cause of the embryonic lethality due to genomic hypomethylation is not clear. Proposed mechanisms include the dysregulation of developmentally expressed genes (BIRD 1997), disrupted host-genome defense (WALSH et al. 1998) and inappropriate X inactivation (BEARD et al. 1995). Despite the fact that the $Dnmt1^n$ allele is a recessive lethal allele, it is not a null allele (LEI et al. 1996). Low levels of alternative splicing allow the generation of functional DNA methyltransferase from this allele (LEI et al. 1996). Two truly non-functional alleles have been generated. One is a $Neo^R$ insertion in the replication foci targeting region (LEI et al. 1996). This allele is referred to as the $Dnmt1^s$ allele for the SalI site used for the insertion (LEI et al. 1996). The other null allele is a deletion of conserved motifs in the C-terminal catalytic domain and is designated the $Dnmt1^c$ allele (LEI et al. 1996). The $Dnmt1^s$ allele and the $Dnmt1^c$ allele are both homozygous recessive mutations showing

**Table 1.** Knockout alleles of the murine *Dnmt1* gene

| *Dnmt1* allele | Approximate functional expression level (% of wild type) | Homozygous phenotype in embryonic stem cells | Homozygous phenotype in vivo | Reference |
| --- | --- | --- | --- | --- |
| $Dnmt1^+$ | 100 | Normal | Normal | (BESTOR et al. 1988) |
| $Dnmt1^p$ or $Dnmt1^{p,DNA}$ | 10–50 | Normal | Normal | (TUCKER et al. 1996a) |
| $Dnmt1^n$ | 5 | Hypomethylated | Lethal at E10.5 | (LI et al. 1992) |
| $Dnmt1^s$ | 0 | Hypomethylated | Lethal at E8.5 | (LEI et al. 1996) |
| $Dnmt1^c$ | 0 | Hypomethylated | Lethal at E8.5 | (LEI et al. 1996) |

extreme loss of DNA methylation and an associated embryonic lethality at about day 8.5. ES cells that are homozygous for any of the severe *Dnmt1* mutations are viable in the undifferentiated state but die upon differentiation in vitro or in vivo (LEI et al. 1996).

An additional modified *Dnmt1* allele has been described in which a hybrid genomic/cDNA *Dnmt1* construct has been inserted into the endogenous *Dnmt1* locus by an insertion-type homologous-recombination event (TUCKER et al. 1996a). The insertion strategy employed in this case is referred to as CHIP (cDNA homologous-insertion protocol). The resulting allele produces functional Dnmt1 transcripts but at levels that may be slightly lower than those of the wild-type locus. If this can be confirmed, then the allele could be useful as a hypomorphic allele in combination with more severe knockout alleles. In our laboratory, we refer to this allele as the *Dnmt1$^P$* allele because of the *Puro$^R$* marker used for the targeting event. *Dnmt1$^{P/P}$*, *Dnmt1$^{n/P}$* and *Dnmt1$^{s/P}$* mice are all fully viable and normal in appearance. It remains to be determined to what extent DNA-methylation levels are affected in these mice.

### 3.3.2 Antisense and Oligonucleotides

Expression of an antisense *Dnmt1* construct in a mouse adrenocortical tumor cell line has been shown to lead to DNA demethylation and inhibition of tumorigenesis in syngeneic mice (MACLEOD and SZYF 1995). The introduction of phosphorothioate-modified antisense oligonucleotides directed against *Dnmt1* had similar effects on this cell line (RAMCHANDANI et al. 1997). Presumably, these antisense nucleic acids operate at the *Dnmt1* RNA-transcript level. However, inhibitory substrate oligonucleotides have also been deployed as direct DNA-methyltransferase inhibitors. Oligonucleotides containing 5-fluorocytosine incorporated in a hemi-methylated CpG context are suicide inhibitors of mammalian DNA methyltransferases (SMITH et al. 1992; YODER et al. 1997; FLYNN and REICH 1998). DNA-methyltransferase antagonists based on phosphorothioate-modified substrate oligonucleotides with secondary structures have been shown to be active on live cells in culture (BIGEY et al. 1999). Other oligonucleotide inhibitors of the Dnmt1 enzyme that appear to rely on allosteric inhibition rather than on occupation of the DNA-substrate-binding site have been developed (Reich, personal communication).

### 3.3.3 Drugs

Inhibitory drugs have been particularly useful tools in DNA methylation research ever since the important discovery that the cytidine analogs 5-azacytidine (5-azaC) and 5-aza-2'-deoxycytidine (5-azaCdR) are potent mechanism-based inhibitors of DNA (cytosine-5)-methyltransferases (JONES and TAYLOR 1980; SANTI et al. 1984). These two drugs have been instrumental in hundreds of DNA-methylation studies in the past two decades. Their specificity is due to the fact that the DNA (cytosine-5)-methyltransferase forms a stable covalent bond with the C-5 position of incorporated 5-azacytosine. It is important to note that the demethylating action of 5-azaC and 5-azaCdR derives from the depletion of free active DNA methyltransferase by this trapping action, as opposed to a mere resistance of the incorporated aza analog to undergoing methylation. Another essential feature of the mechanism is that 5-azaC and 5-azaCdR are essentially prodrugs that only become active inhibitors upon incorporation into DNA. The 5-fluorocytidine and 5-fluoro-2'-deoxycytidine analogs have a mode of action similar to those of the aza analogs but are more toxic (JONES and TAYLOR 1980; SANTI et al. 1983). It is often assumed that the toxicity associated with these drugs is attributable to the loss of genomic-DNA methylation. However, the irreversible covalent bond formed between the incorporated drug and the DNA (cytosine-5)-methyltransferase enzyme is perhaps a much more important factor in the toxicity (MICHALOWSKY and JONES 1987). Indeed, studies in mice have shown that the drug is more toxic to wild-type mice than it is to $Dnmt1^{n/+}$ heterozygous mice, which presumably have lower DNA (cytosine-5)-methyltransferase expression and less DNA methylation (JUTTERMANN et al. 1994). This result is more consistent with the enzyme–drug interaction (rather than DNA hypomethylation) being the major source of toxicity (JUTTERMANN et al. 1994). The significance of the covalent enzyme–drug complex is also evident from studies of 5-azaCdR mutagenicity (JACKSON-GRUSBY et al. 1997). Mutations arising at CpG dinucleotides are specifically increased and predominate in mice treated with 5-azaCdR despite the fact that the majority of incorporated analogs would be expected to occur outside of a CpG-sequence context. This implicates the DNA-methyltransferase enzyme, which recognizes CpG as its target sequence.

DNA-methyltransferase inhibitors have been powerful research tools, and it is anticipated that they will be of clinical interest as well. Although 5-azaCdR is currently being evaluated as a cancer-chemotherapeutic agent, its cytotoxicity may be more attributable to complex formation with DNA methyltransferase than to DNA hypomethylation. Drugs that inhibit DNA methyltransferases before they interact with the genomic DNA would be expected to achieve a better ratio of DNA-methylation inhibition to toxicity. SAM analogs, such as sinefungin, have been explored for this purpose, but they affect other transmethylation reactions in the cell and may lead to increased DNA-methyltransferase-induced cytosine deamination (ZINGG et al. 1996). The substrate-inhibition oligonucleotides described in the previous section are costly and inefficient, but they are representative of the desired mode of action.

# 4 Applications

## 4.1 DNA-Methyltransferase Activities in ES Cells

Gene-targeting technology has been particularly useful in the analysis of mammalian DNA methyltransferases. The conserved motifs I–VIII in the catalytic domain of *Dnmt1* are homozygously deleted in $Dnmt1^{c/c}$ ES cells. Nevertheless, not only do these cells have low levels of residual DNA methylation, they are also able to methylate newly introduced retroviral genomes with initial kinetics similar to those for wild-type ES cells (LEI et al. 1996). Since all DNA (cytosine-5)-methyl-transferases identified so far in nature contain the conserved motifs I, IV, VI and VIII, this was considered good evidence that there must be at least one other gene encoding a DNA methyltransferase that contains these motifs and is expressed in ES cells. The *Dnmt3a* and *Dnmt3b* genes are considered good candidates for this role.

If the residual DNA methylation in *Dnmt1* homozygous knockout ES cells is attributable to the activity of a different enzyme, then the distribution pattern of this residual DNA methylation should reflect the sequence preference of this other enzyme. However, no shift in the relative distribution of 5-methylcytosine in mouse A repeats was found in $Dnmt1^{n/n}$ and $Dnmt1^{s/s}$ ES cells compared with wild-type ES cells, although the levels of methylation were lower (WOODCOCK et al. 1998). The putative other DNA methyltransferase may have a sequence preference similar to that of Dnmt1, but more sites in the genome need to be analyzed before coming to that conclusion.

## 4.2 *Dnmt1* Knockout Mice in Cancer Research

The in vivo modulation of DNA methylation offers the opportunity to probe causal relationships between DNA methylation and cancer. Genetic manipulation of DNA-methyltransferase genes provides a means of accomplishing this with high precision. However, homozygous knockout *Dnmt1* mice are not viable, which prevents their use in the assessment of DNA methylation on postnatal events, such as oncogenesis. This dilemma was partially resolved by combining *Dnmt1* hetero-zygosity with low doses of 5-azaCdR (LAIRD et al. 1995). In this particular model system, tissues of $Dnmt1^{S/+}$ mice showed slight DNA hypomethylation compared with wild-type mice. $Dnmt1^{+/+}$ mice treated weekly with 5-azaCdR were more noticeably hypomethylated, while the combination of $Dnmt1^{S/+}$ and 5-azaCdR caused substantial hypomethylation (LAIRD et al. 1995). This in vivo modulation system was applied to the Min-mouse model of intestinal neoplasia. Min mice carry a germline mutation of the *Apc* tumor-suppressor gene and are consequently pre-disposed to the development of intestinal adenomas. This system was used to test the effects of DNA hypomethylation on polyp multiplicity (LAIRD et al. 1995). $Dnmt1^{+/+} Apc^{Min/+}$ developed an average of 113 intestinal adenomas, while $Dnmt1^{S/+} Apc^{Min/+}$ mice developed an average of 46 adenomas. $Dnmt1^{+/+} Apc^{Min/+}$ mice treated with 5-azaCdR showed a further reduction to only 20 adenomas, while $Dnmt1^{S/+} Apc^{Min/+}$ mice were virtually polyp-free, with an average of only two intestinal adenomas. These results not only argue for a causal involvement of DNA methylation in intestinal polyp formation, they also suggest that sufficient levels of DNA methylation may even be a prerequisite for the development of adenomas.

Several molecular mechanisms have been proposed to explain the suppression of polyp multiplicity by DNA hypomethylation (LAIRD et al. 1995). Reduced numbers of polyps in mice with DNA hypomethylation argues against the activation of proto-oncogenes by DNA hypomethylation as a rate-determining step in polyp formation (COUNTS and GOODMAN 1995). A reduction of methylation-dependent CpG transition mutations seems unlikely in light of the increased rate of mutagenesis observed in 5-azaCdR-treated mice (JACKSON-GRUSBY et al. 1997). Likewise, the increased genomic instability observed in homozygous *Dnmt1*-knockout ES cells suggests that deletion events are not the rate-limiting step influenced by DNA methylation in polyp formation in Min mice (CHEN et al. 1998). One plausible mechanism is that early polyp development depends on transcriptional silencing of growth-suppressor genes by DNA hypermethylation (JONES and LAIRD 1999). Mice with reduced levels of active Dnmt1 would be less capable of growth-suppressor-gene silencing. This is one of the most difficult mechanisms to test experimentally. We are currently investigating whether CpG-island hypermethylation does indeed occur during intestinal polyp formation in Min mice and, if so, whether it is less frequent or extensive in mice with reduced levels of active Dnmt1 enzyme.

The tumor-suppressive effect of DNA hypomethylation seen in Min mice is not necessarily applicable to other types of cancer. Indeed, 5-aza-CdR treatment increases the incidence of T cell lymphomagenesis in $Dnmt1^{S/+}$ mice to a much greater extent than in wild-type mice (JACKSON-GRUSBY and JAENISCH 1996). This suggests that DNA hypomethylation increases the rate of lymphomagenesis. Other studies have shown that 5-azaC treatment can increase the incidence of lung tumors and granulocytic sarcomas (Carcinogenesis-Testing Program 1978; STONER et al. 1973). A caveat to these studies is that the mutagenicity of 5-azaC (JACKSON-GRUSBY et al. 1997) could have been responsible for the observed increase in tumorigenesis. The inclusion of *Dnmt1*-heterozygous mice in a study allows the discrimination between a mutagenic mechanism and a hypomethylation mechanism of action for 5-azaC. Enzyme-mediated mutagenesis would be expected to be less severe in *Dnmt1* heterozygotes, whereas DNA hypomethylation should be more severe in the heterozygotes. *Dnmt1*-knockout mice will be useful for the dissection of the role of DNA methylation in various tumor models yet to be investigated.

# 5 Outlook

Although *Dnmt1*-knockout mice have been informative in the Min-mouse model, it is clear that the non-viability of the homozygous mutant mice remains a major impediment to the use of these mice in cancer research. The obvious next step will be the generation of conditional knockouts in which the loss of function of Dnmt1 is limited temporally or spatially during development or in adult tissues. Efforts to develop such cell-type or developmentally controlled knockouts are underway in several laboratories. The most successful system used to generate tissue-specific knockouts has been the Cre–lox system, in which tissue-restricted expression of the Cre recombinase mediates lox-dependent deletion of one or more essential exons of a gene in a tissue-specific fashion. A conceptual problem with the application of this system to the *Dnmt1* gene is that there is evidence that *Dnmt1*-homozygous mutations are growth inhibitory or even cell lethal in differentiated cell types (LEI et al. 1996). Thus, differentiated adult cells undergoing Cre-mediated knockout of the *Dnmt1* gene might merely die, preventing the assessment of severe hypomethylation on tumor processes in this cell type. As an alternative, my laboratory is developing tissue-specific transcriptional repression of the endogenous *Dnmt1* gene using a prokaryotic operator/repressor system. It remains to be seen, however, whether this system can achieve sufficient levels of transcriptional repression of the endogenous *Dnmt1* gene.

Yet another approach to the development of tissue-restricted knockout of the *Dnmt1* gene would be the development of transgenic mouse strains that express the *Dnmt1* gene at appropriate levels in those tissues necessary for viability but not in the adult tissue of interest. Such transgenic lines could then be crossed with the constitutive *Dnmt1*-knockout strains to create mice with a methylation defect restricted to particular tissues. The development of a transgenic line with such precise expression requirements is a formidable technical challenge, however.

Now that more insight has been gained into the structure of upstream exons of the *Dnmt1* gene, new attempts at generating *Dnmt1*-transgenic mice should be undertaken. Mild overexpression of DNA methyltransferases would be useful tools in elucidating the effects of DNA hypermethylation in vivo.

The identification of *Dnmt2, Dnmt3a, Dnmt3b* and the demethylase gene provide new targets for knockout and transgenic approaches. Initial analysis suggests that the *Dnmt3* genes are not upregulated in human colorectal adenocarcinomas (EADS et al. 1999). Nevertheless, they remain interesting candidates for a role in contributing to the abnormal DNA methylation patterns seen in cancer cells. Elucidation of the regulatory elements that control the distribution pattern of 5-methylcytosine in the genome is essential to our understanding of how these patterns become disrupted in cancer cells. Mouse models will allow us to test their contribution to oncogenesis once the elements have been defined.

*Acknowledgements.* Research conducted by the author's laboratory is supported by National Institutes of Health/National Cancer Institute grant R01 CA 75090.

# References

Beard C, Li E, Jaenisch R (1995) Loss of methylation activates *Xist* in somatic but not in embryonic cells. Genes Dev 9:2325–2334
Bestor TH (1992) Activation of mammalian DNA methyltransferase by cleavage of a Zn-binding regulatory domain. Embo J 11:2611–7

Bestor TH, Laudano A, Mattaliano R, Ingram V (1988) Cloning and sequencing of a cDNA encoding DNA methyltransferase of mouse cells. The carboxyl-terminal domain of the mammalian enzymes is related to bacterial restriction methyltransferases. J Mol Biol 203:971–983

Bhattacharya SK, Ramchandani S, Cervoni N, Szyf M (1999) A mammalian protein with specific demethylase activity for mCpG DNA. Nature 397:579–583

Bigey P, Knox JD, Croteau S, Bhattacharya SK, Theberge J, Szyf M (1999) Modified oligonucleotides as bona fide antagonists of proteins interacting with DNA. Hairpin antagonists of the human DNA methyltransferase. J Biol Chem 274:4594–4606

Bird A (1997) Does DNA methylation control transposition of selfish elements in the germline? Trends Genet 13:469–472

Carcinogenesis-Testing-Program NCI (1978) Bioassay of 5-aza-cytidine for possible carcinogenicity. NCI Carcinogenesis Technical Report Series 42:1–86 CAS No 320-67-2

Cedar H, Verdine GL (1999) Gene expression. The amazing demethylase. Nature 397:568–569

Chen J, Maxson R, Jones PA (1993) Direct induction of DNA hypermethylation in sea urchin embryos by microinjection of 5-methyl dCTP stimulates early histone gene expression and leads to developmental arrest. Dev Biol 155:75–86

Chen RZ, Pettersson U, Beard C, Jackson-Grusby L, Jaenisch R (1998) DNA hypomethylation leads to elevated mutation rates. Nature 395:89–93

Chuang LS, Ian HI, Koh TW, Ng HH, Xu G, Li BF (1997) Human DNA-(cytosine-5) methyltransferase-PCNA complex as a target for p21 WAF1. Science 277:1996–2000

Counts JL, Goodman JI (1995) Alterations in DNA methylation may play a variety of roles in carcinogenesis. Cell 83:13–15

Counts JL, McClain RM, Goodman JI (1997) Comparison of effect of tumor promoter treatments on DNA methylation status and gene expression in B6C3F1 and C57BL/6 mouse liver and in B6C3F1 mouse liver tumors. Mol Carcinog 18:97–106

Eads CA, Danenberg KD, Kawakami K, Saltz LB, Danenberg PV, Laird PW (1999) CpG island hypermethylation in human colorectal tumors is not associated with DNA methyltransferase overexpression. Cancer Research, In Press

Flynn J, Reich N (1998) Murine DNA (cytosine-5)-methyltransferase: steady-state and substrate trapping analyses of the kinetic mechanism. Biochemistry 37:15162–15169

Glickman JF, Pavlovich JG, Reich NO (1997) Peptide mapping of the murine DNA methyltransferase reveals a major phosphorylation site and the start of translation. J Biol Chem 272:17851–17857

Holliday R, Ho T (1998) Evidence for gene silencing by endogenous DNA methylation. Proc Natl Acad Sci USA 95:8727–8732

Jackson-Grusby L, Jaenisch R (1996) Experimental manipulation of genomic methylation. Semin Cancer Biol 7:261–268

Jackson-Grusby L, Laird PW, Magge SN, Moeller BJ, Jaenisch R (1997) Mutagenicity of 5-aza-2'-deoxycytidine is mediated by the mammalian DNA methyltransferase. Proc Natl Acad Sci USA 94:4681–4685

Jones PA, Laird PW (1999) Cancer epigenetics comes of age. Nat Genet 21:163–167

Jones PA, Taylor SM (1980) Cellular differentiation, cytidine analogs and DNA methylation. Cell 20: 85–93

Jost JP, Fremont M, Siegmann M, Hofsteenge J (1997) The RNA moiety of chick embryo 5-methylcytosine- DNA glycosylase targets DNA demethylation. Nucleic Acids Res 25:4545–4550

Juttermann R, Li E, Jaenisch R (1994) Toxicity of 5-aza-2'-deoxycytidine to mammalian cells is mediated primarily by covalent trapping of DNA methyltransferase rather than DNA demethylation. Proc Natl Acad Sci USA 91:11797–11801

Laird PW, Jaenisch R (1994) DNA methylation and cancer. Hum Mol Genet 3:1487–95

Laird PW, Jackson-Grusby L, Fazeli A, Dickinson SL, Jung WE, Li E, Weinberg RA, Jaenisch R (1995) Suppression of intestinal neoplasia by DNA hypomethylation. Cell 81:197–205

Lei H, Oh SP, Okano M, Juttermann R, Goss KA, Jaenisch R, Li E (1996) De novo DNA cytosine methyltransferase activities in mouse embryonic stem cells. Development 122:3195–3205

Leonhardt H, Page AW, Weier HU, Bestor TH (1992) A targeting sequence directs DNA methyltransferase to sites of DNA replication in mammalian nuclei. Cell 71:865–73

Lertratanangkoon K, Orkiszewski RS, Scimeca JM (1996) Methyl-donor deficiency due to chemically induced glutathione depletion. Cancer Res 56:995–1005

Lertratanangkoon K, Wu CJ, Savaraj N, Thomas ML (1997) Alterations of DNA methylation by glutathione depletion. Cancer Lett 120:149–156

Li E, Bestor TH, Jaenisch R (1992) Targeted mutation of the DNA methyltransferase gene results in embryonic lethality. Cell 69:915–926

Liu Y, Oakeley EJ, Sun L, Jost JP (1998) Multiple domains are involved in the targeting of the mouse DNA methyltransferase to the DNA replication foci. Nucleic Acids Res 26:1038–1045

MacLeod AR, Szyf M (1995) Expression of antisense to DNA methyltransferase mRNA induces DNA demethylation and inhibits tumorigenesis. J Biol Chem 270:8037–8043

Mertineit C, Yoder JA, Taketo T, Laird DW, Trasler JM, Bestor TH (1998) Sex-specific exons control DNA methyltransferase in mammalian germ cells. Development 125:889–897

Michalowsky LA, Jones PA (1987) Differential nuclear protein binding to 5-azacytosine-containing DNA as a potential mechanism for 5-aza-2'-deoxycytidine resistance. Mol Cell Biol 7:3076–83

Migchielsen AA, Breuer ML, Van Roon MA, te Riele H, Zurcher C, Ossendorp F, Toutain S, Hershfield MS, Berns A, Valerio D (1995) Adenosine-deaminase-deficient mice die perinatally and exhibit liver-cell degeneration, atelectasis and small intestinal cell death. Nat Genet 10:279–287

Miller MW, Duhl DM, Winkes BM, Arredondo-Vega F, Saxon PJ, Wolff GL, Epstein CJ, Hershfield MS, Barsh GS (1994) The mouse lethal nonagouti (a(x)) mutation deletes the S-adenosylhomocysteine hydrolase (Ahcy) gene. Embo J 13:1806–16

Nyce JW (1997) Drug-induced DNA hypermethylation: a potential mediator of acquired drug resistance during cancer chemotherapy. Mutat Res 386:153–161

Okano M, Xie S, Li E (1998a) Cloning and characterization of a family of novel mammalian DNA (cytosine-5) methyltransferases. Nature Genet 19:219–220

Okano M, Xie S, Li E (1998b) Dnmt2 is not required for de novo and maintenance methylation of viral DNA in embryonic stem cells. Nucleic Acids Res 26:2536–2540

Pinarbasi E, Elliott J, Hornby DP (1996) Activation of a yeast pseudo DNA methyltransferase by deletion of a single amino acid. J Mol Biol 257:804–813

Pogribny IP, Basnakian AG, Miller BJ, Lopatina NG, Poirier LA, James SJ (1995) Breaks in genomic DNA and within the p53 gene are associated with hypomethylation in livers of folate/methyl-deficient rats. Cancer Res 55:1894–901

Pogribny IP, Miller BJ, James SJ (1997) Alterations in hepatic p53 gene methylation patterns during tumor progression with folate/methyl deficiency in the rat. Cancer Lett 115:31–38

Poirier LA (1994) Methyl group deficiency in hepatocarcinogenesis. Drug Metab Rev 26:185–99

Ramchandani S, MacLeod AR, Pinard M, von Hofe E, Szyf M (1997) Inhibition of tumorigenesis by a cytosine-DNA methyltransferase antisense oligodeoxynucleotide. Proc Natl Acad Sci USA 94:684–689

Robertson KD, Uzvolgyi E, Liang G, Talmadge C, Sumegi J, Gonzales FA, Jones PA (1999) The human DNA methyltransferases (DNMTs) 1, 3a and 3b: coordinate mRNA expression in normal tissues and overexpression in tumors. Nucleic Acids Res 27:2291–2298

Santi DV, Garrett CE, Barr PJ (1983) On the mechanism of inhibition of DNA-cytosine methyltransferases by cytosine analogs. Cell 33:9–10

Santi DV, Norment A, Garrett CE (1984) Covalent-bond formation between a DNA-cytosine methyltransferase and DNA containing 5-azacytosine. Proc Natl Acad Sci USA 81:6993–7

Shen JC, Rideout Wd, Jones PA (1992) High frequency mutagenesis by a DNA methyltransferase. Cell 71:1073–80

Smith SS, Kaplan BE, Sowers LC, Newman EM (1992) Mechanism of human methyl-directed DNA methyltransferase and the fidelity of cytosine methylation. Proc Natl Acad Sci USA 89:4744–4748

Stoner GD, Shimkin MB, Kniazeff AJ, Weisburger JH, Weisburger EK, Gori GB (1973) Test for carcinogenicity of food additives and chemotherapeutic agents by the pulmonary response in strain A mice. Cancer Res 33:3069–85

Swisher JF, Rand E, Cedar H, Marie Pyle A (1998) Analysis of putative RNase sensitivity and protease insensitivity of demethylation activity in extracts from rat myoblasts. Nucleic Acids Res 26:5573–5580

Tucker KL, Beard C, Dausmann J, Jackson-Grusby L, Laird PW, Lei H, Li E, Jaenisch R (1996a) Germ-line passage is required for establishment of methylation and expression patterns of imprinted but not of non-imprinted genes. Genes Dev 10:1008–1020

Tucker KL, Talbot D, Lee MA, Leonhardt H, Jaenisch R (1996b) Complementation of methylation deficiency in embryonic stem cells by a DNA methyltransferase minigene. Proc Natl Acad Sci USA 93:12920–12925

Van den Wyngaert I, Sprengel J, Kass SU, Luyten WH (1998) Cloning and analysis of a novel human putative DNA methyltransferase. FEBS Lett 426:283–289

Vertino PM, Yen RW, Gao J, Baylin SB (1996) De novo methylation of CpG-island sequences in human fibroblasts overexpressing DNA (cytosine-5)-methyltransferase. Mol Cell Biol 16:4555–4565

Walsh CP, Chaillet JR, Bestor TH (1998) Transcription of IAP endogenous retroviruses is constrained by cytosine methylation. Nat Genet 20:116–117

Weiss A, Keshet I, Razin A, Cedar H (1996) DNA demethylation in vitro: involvement of RNA. Cell 86:709–718

Wilkinson CR, Bartlett R, Nurse P, Bird AP (1995) The fission yeast gene *pmt1*+ encodes a DNA-methyltransferase homologue. Nucleic Acids Res 23:203–10

Woodcock DM, Linsenmeyer ME, Warren WD (1998) DNA methylation in mouse A-repeats in DNA methyltransferase-knockout ES cells and in normal cells determined by bisulfite genomic sequencing. Gene 206:63–67

Wu J, Issa JP, Herman J, Bassett DE, Jr Nelkin BD, Baylin SB (1993) Expression of an exogenous eukaryotic DNA methyltransferase gene induces transformation of NIH 3T3 cells. Proc Natl Acad Sci USA 90:8891–8895

Wu J, Herman JG, Wilson G, Lee RY, Yen RW, Mabry M, de Bustros A, Nelkin BD, Baylin SB (1996) Expression of prokaryotic HhaI DNA methyltransferase is transforming and lethal to NIH 3T3 cells. Cancer Res 56:616–622

Xu GL, Bestor TH (1997) Cytosine methylation targetted to pre-determined sequences. Nat Genet 17:376–378

Yoder JA, Yen RWC, Vertino PM, Bestor TH, Baylin SB (1996) New 5' regions of the murine and human genes for DNA (cytosine-5)- methyltransferase. J Biol Chem 271:31092–31097

Yoder JA, Soman NS, Verdine GL, Bestor TH (1997) DNA (cytosine-5)-methyltransferases in mouse cells and tissues. Studies with a mechanism-based probe. J Mol Biol 270:385–395

Yoder JA, Bestor TH (1998) A candidate mammalian DNA methyltransferase related to pmt1p of fission yeast. Hum Mol Genet 7:279–284

Zingg JM, Shen JC, Yang AS, Rapoport H, Jones PA (1996) Methylation inhibitors can increase the rate of cytosine deamination by (cytosine-5)-DNA methyltransferase. Nucleic Acids Res 24:3267–3275

# DNA Methylation Inhibitors in the Treatment of Leukemias, Myelodysplastic Syndromes and Hemoglobinopathies: Clinical Results and Possible Mechanisms of Action

M. LÜBBERT

## 1 Introduction

As recently reviewed (BAYLIN et al. 1998), cytosine hypermethylation of numerous genes important in orderly cell proliferation and maturation (many of them *bone fide* or putative tumor-suppressor genes) is frequent in primary neoplasias and tumor cell lines (HERMAN et al. 1995; ZINGG and JONES 1997). Therefore, the application of pharmacologic inhibitors of DNA methylation provides a conceptually

Department of Medicine, Division of Hematology/Oncology, University of Freiburg Medical Center, Hugstetter Str. 55, D-79106 Freiburg, Germany
E-mail: luebbert@mm11.ukl.uni-freiburg.de

attractive and rational approach to reverting these epigenetic changes in the malignant clone and re-establish the antiproliferative and possibly differentiation-inducing signals silenced by hypermethylation.

5-Azacytidine (azacitidine) and its deoxy congener 5-aza-2'-deoxycytidine (decitabine) have been clinically developed based on their strong in vitro and in vivo antileukemic activity at cytotoxic concentrations, and their differentiation-inducing potential at lower concentrations in cell line models of hematopoietic (CHRISTMAN et al. 1983; MOMPARLER et al. 1985; VISVADER and ADAMS 1993) and non-hematopoietic lineages (TAYLOR and JONES 1982; CREMISI 1989). In relapsed and refractory acute leukemia, both drugs show a clinical activity (in terms of response rates and response duration) that is comparable to those of other "second line" agents. Their mechanism of action, however, is probably distinct from those of other pyrimidine antimetabolites, including Ara-C. This is suggested by the markedly prolonged and delayed leukemic blast lysis frequently observed in patients with acute leukemias treated with these drugs, and by the recurrent lack of severe bone marrow hypoplasia preceding the hematologic responses to low-dose azacitidine or decitabine in myelodysplasia.

Azacitidine and decitabine were both synthesized in 1964 by Sorm and co-workers (PISKALA and SORM 1964; PLIML and SORM 1964). Both are ring analogues of the pyrimidine nucleosides cytidine and 2'-deoxycytidine, respectively, but differ from these by a nitrogen at the fifth carbon position. The primary activation of azacitidine is by uridine–cytidine kinase, whereas decitabine is activated by deoxycytidine kinase. Following their phosphorylation to monophosphates, both compounds are incorporated into newly synthesized DNA. Azacitidine, being a ribonucleoside, is predominantly incorporated into newly synthesized RNA. Inactivation of both drugs occurs through deamination by cytidine deaminase. Chemically, both are quite unstable in aqueous solutions, making frequent changes of newly prepared solutions necessary.

Following incorporation of azacitidine and decitabine into DNA, newly hypomethylated DNA strands are synthesized (JONES and TAYLOR 1981; WILSON et al. 1983). Both compounds act as inhibitors of the major DNA methyltransferase (now termed Dnmt1) by forming a covalent complex with this enzyme (BOUCHARD and MOMPARLER 1983; SANTI et al. 1983). At equimolar concentrations, decitabine is at least twice as potent as azacitidine in inhibiting DNA methylation (CREUSOT et al. 1981).

In vitro studies using cell lines have shown a time- and dose-dependent inhibition of proliferation of both compounds. At high concentrations, there is a cytotoxic effect, which may be related in part to the synthesis of alkali-labile DNA strands (D'INCALCI et al. 1985). At lower concentrations, both act as weak inducers of differentiation of myeloid leukemic cell lines (CREUSOT et al. 1981; CHRISTMAN et al. 1983; MOMPARLER et al. 1985). Primary leukemic myeloid cells were shown by PINTO et al. (1984a) to have a propensity for in vitro granulocytic or monocytic differentiation induced by decitabine. In mouse models, both azacitidine (SORM and VESELY 1964; LI et al. 1974) and decitabine (WILSON et al. 1983) have marked antileukemic activity. In a rat model for carcinogenicity, azacitidine behaved as a complete carcinogen, whereas decitabine and four other cytidine analogues, when tested in a limited series of experiments, did not (CARR et al. 1988). However, because of this effect (in the L1210 murine leukemia and for lymphoid leukemia in AK mice) in combination with their demethylating potential, both were interesting candidates for clinical studies in patients with acute leukemias.

# 2 Development of Azacitidine for Remission Induction and Post-Remission Treatment of Acute Myeloid Leukemia

Azacitidine administered intramuscularly was first shown by HRODEK and VESELY (1971) to have some antileukemic activity in childhood acute lymphoblastic leukemia (ALL), inducing a decrease in peripheral blood leukocytes in the absence of additional corticosteroids. In parallel with this encouraging report, phase-I studies to determine the toxicity profile and maximally tolerated dose of this drug were initiated in the United States. These were reviewed by VON HOFF et al. (1976). The maximally tolerated dosages of azacitidine were between $300mg/m^2$ body surface and $1125mg/m^2$, with schedules of drug administration varying between 1 day and 10 days. The major non-hematologic toxicities were nausea and vomiting (particularly with bolus injection), diarrhea and, less frequently, hepatic toxicity. The major hematologic toxicities were neutropenia and thrombocytopenia, which were most pronounced in patients with hematologic malignancies (VOGLER et al. 1976).

The role of azacitidine as a single agent in the treatment of acute myeloid leukemia (AML) was examined in eight phase-II studies, which included 200 patients between 1973 and 1976. These studies were also reviewed by VON HOFF et al. (1976). Almost all patients had previously been treated with cytosine arabinoside (Ara-C)-based standard protocols. An overall response rate of 36% was observed, with 20% complete remissions (CR) and 16% partial remissions (PR), with a median duration of 17 weeks. Since these results clearly demonstrated activity of azacitidine in relapsed and refractory AML, combinations of azacitidine with other drugs were explored. As reviewed by Von Hoff and coworkers, a total of 66 patients were treated in five studies (combinations with daunorubicin, cytarabine, vincristin and other drugs), resulting in complete response rates ranging between 29% and 58%. However, the numbers of patients treated in each study were too small to allow conclusions regarding the comparable efficacy of the various combinations. The toxicity profile observed in phase-I/II studies of azacitidine included nausea and vomiting in 73%, diarrhea in 53% and dose-related myelosuppression, which does not appear to be cumulative in patients with solid tumors. Infrequent toxicities included abnormal findings in liver-function tests (7%) and a peculiar myalgic/asthenic syndrome (generalized muscle tenderness, weakness and lethargy) in 2.7% of patients (VON HOFF et al. 1976).

GLOVER et al. (1987) expanded these data in an updated review of all clinical studies performed with azacitidine and sponsored by the National Cancer Institute up to 1985. Of 335 patients with AML (two-thirds of them relapsed or refractory to previous treatment) treated in 11 different studies using azacitidine alone, the overall complete response rate had decreased from 20% to 16.7% (GLOVER et al. 1987). In the largest of these studies (SAIKI et al. 1981), only eight complete responses were observed in 101 patients.

More recently, a small study aimed at exploring a low-dose approach of azacitidine in relapsed or secondary AML was conducted (LEE et al. 1990). Azacitidine was administered by continuous infusion at $75mg/m^2/day$ for 7 days to 11 patients with AML (median age 55 years). Twenty-one courses were administered without significant non-hematologic toxicity. The majority of patients (7 of 11) received two complete courses of treatment. No objective hematologic response or clinical benefit was observed.

Studies using azacitidine in combination with a second compound for re-induction of remission in relapsed/refractory AML were also systematically reviewed by GLOVER et al. (1987). However, sample size in the 15 studies evaluated (median of 16 patients; range 6–81) again did not allow comparison of the relative efficacy of these combinations. The largest study was performed with 81 adult patients with relapsed AML (treated with $150mg/m^2$ azacitidine daily as a continuous infusion for 5 days and $300mg/m^2$ deoxythioguanosine for 5 days). In this study, 21 objective responses were observed (16 complete and 5 partial responses; OMURA et al. 1979). Among the other trials, the most frequently studied two-drug combination of azacitidine in adults was with amsacrine ($150mg/m^2$ for 5 days). In pediatric hematology, the combination of azacitidine with VP-16 was the subject of the majority of studies.

Among the multidrug trials for leukemia which had azacitidine as one of the drugs, the combination of azacitidine with daunorubicin, Ara-C, prednisone and vincristine ("D-ZAPO") was clearly the most promising regimen (BAEHNER et al. 1979). CRs were observed in 117 of 163 children with untreated AML, making this protocol the only drug combination that includes a DNA methylation inhibitor in *frontline* treatment of AML in a large study. However, due to the single-arm study design, no conclusion as to the contribution of azacitidine to the overall efficacy of this protocol could be drawn.

As of 1987, 16 consolidation and/or maintenance regimens administered following the achievement of CR in AML were active (VOGLER et al. 1984; GLOVER et al. 1987). However, none of these trials was designed to address the role of azacitidine as an individual agent. A long-term follow-up report of a randomized trial comparing three different post-remission regimens was recently published (VOGLER et al. 1995). Of three arms varying in drug combination and overall dose intensity, one arm designed to use non-crossresistant drugs included a 5-day continuous infusion of azacitidine ($150mg/m^2/day$) preceded by 5 days of amsacrine ($120mg/m^2/day$). However, the 31% 5-year survival of patients treated in this arm was not superior to that of the other two arms, which did not contain azacitidine. Several studies testing different drug combinations including azacitidine during post-remission treatment of AML were also performed in Europe (JEHN et al. 1990; JEHN 1994). Neither of these studies was designed to determine the individual role of azacitidine in this setting.

GLOVER and coworkers (1987) pointed out a lack of systematic progression in the development of azacitidine during the period from 1976 to 1987. With one exception, no first-line treatment schedule of leukemia included azacitidine and the role of this drug in consolidation, maintenance and intensification regimens had not been defined. Since then, the third goal formulated by them, i.e., the subcutaneous route of administration, has been systematically studied (see below).

# 3 Activity of Decitabine in the Remission-Induction Treatment of AML

Decitabine was first used by RIVARD and colleagues (1981) as a single agent in a phase-I study examining its activity in the treatment of relapsed or refractory acute leukemia of childhood, in addition to its toxicity profile. Fifteen children with ALL, seven with AML and three with solid tumors were included in the study. The starting doses were 0.75mg to 30mg/kg decitabine administered as 30-h infusions repeated 12 times, or an intravenous (i.v.) bolus administration of 10mg/kg. At these doses, only a short antileukemic effect was noted. However, in a dose-range of 36–80mg/kg infused over 36–44 h, two of nine patients had clearance of bone-marrow blasts below 5%, and one additional patient had reduction of blasts in the bone marrow to less than 20%. In six of the remaining patients, antileukemic activity was limited to reduction of peripheral blood blasts and, in two patients with ALL, reduction of meningeal leukemia cells was noted. Thus, decitabine in these pretreated patients had a dose-dependent antileukemic activity. Responses in the two patients with bone marrow blast suppression were short, with relapses occurring 87 days and 104 days after the first treatment, respectively (Table 1).

No severe acute toxicity was observed up to the highest doses. Mild nausea and vomiting occurred during 11 of 29 treatment courses at the two highest dose levels; at these levels, diarrhea occurred in only four of 29 courses. Alopecia occurred in patients who received more than one treatment at the 30–60mg/kg dose levels or after one treatment with 80mg/kg ($n = 5$). Hematologic toxicity was difficult to evaluate because of massive bone marrow infiltration by leukemia and heavy pretreatment in most of the patients.

In a continuation of this study, MOMPARLER et al. (1985b) treated 27 patients with acute leukemia (21 ALL, 6 AML) with a 36- to 60-h continuous infusion at a total dose of 37–81mg/kg. The majority of children were younger than 10 years. In six patients (22%), a CR was induced; in four others (15%) a PR was achieved, resulting in an overall objective response rate of 37% (33% in ALL patients, 50% in AML patients). Antileukemic activity was present in about half of the remaining patients, suggesting clinical activity in 89% of all patients at these doses. Again, myelotoxicity was the major toxic effect. One patient died of cerebral hemorrhage. Non-hematologic toxicity was moderate (grade 2–3 nausea and vomiting, grade 2 mucositis), and diarrhea was rare.

**Table 1.** Phase-I and -II trials of 5-aza-2'-deoxycytidine (decitabine) in acute leukemia and chronic myelogenous leukemia

| Patients | n | Median age (range) (years) | Drug schedule | Percentage total responses (% CR) | Reference |
|---|---|---|---|---|---|
| relapsed/refractory AML, ALL | 22 (7, 15) | 9 (2–15) | DAC 0.75–80mg/kg (12–44 h c.i.v.) | 14 (9) | RIVARD et al. 1981 |
| relapsed/refractory AML, ALL | 27 (6, 21) | NA (1–20) | DAC 37–80mg/kg (36–60 h c.i.v.) | 37 (22) | MOMPARLER et al. 1993 |
| relapsed/refractory, AML, ALL CML myeloid blast crisis | 26 (21, 2, 3) | 55 (24–74) | DAC 300–500mg/m$^2$ (24–120 h c.i.v.) | 26 (4) | DEBUSSCHER et al. 1990 |
| De novo AML, AML from MDS, CML (not previously treated) | 22 (12, 8, 2) | 74 (62–83) | DAC 45–180mg/m$^2$/ day × 3 (i.v.) ( = 15–60mg/m$^2$/ 4 h t.i.d) | 68 (11) | ZAGONEL et al. 1990 |
| AML (not previously treated) | 12 | 64 (47–77) | DAC 270–360mg/m$^2$/ day × 3 (i.v.) ( = 90–120mg/m$^2$/ 4 h t.i.d) | 40 (30) | PETTI et al. 1993 |
| AML (relapsed/refractory) | 5 | 26 (22–48) | DAC 250–500 mg/m$^2$ b.i.d. × 6–12 | 20 (0) | RICHEL et al. 1991 |
| AML (relapsed/refractory) | 11 | 51 (20–63) | DAC 125–250mg/m$^2$ b.i.d. × 12, amsacrine 120mg/ m$^2$/day × 2 | 82 (73) | RICHEL et al. 1991 |
| AML (relapsed/refractory) | 49 | 52 (18–65) | DAC 125–250mg/m$^2$ b.i.d. × 12, amsacrine 120mg/ m$^2$/day × 2 | 41 (41) | WILLEMZE et al. 1993, 1997 |
| AML (relapsed/refractory) | 36 | 57 (NA) | DAC 125–250mg/m$^2$ b.i.d. × 12, idarubicin 12mg/ m$^2$/day × 3 | 44 (44) | WILLEMZE et al. 1993, 1997 |
| De novo AML (not previously treated) | 8 | 44 (30–59) | DAC 90mg/m$^2$/day × 5, daunorubicin 50mg/m$^2$/day × 3 | 100 (100) | SCHWARTSMANN et al. 1997 |
| CML (myeloid blast crisis) | 31 | 52 (23–78) | DAC 50–100mg/m$^2$ b.i.d. × 10 | 84 (10) | KANTARJIAN et al. 1997a, 1997b; SACCHI et al. 1998 |
| CML (accelerated phase) | 17 | NA | DAC 50–100mg/ m$^2$ b.i.d. × 10 | 53 (0) | KANTARJIAN et al. 1997a, 1997b |

*ALL*, acute lymphoblastic leukemia; *AML*, acute myeloid leukemia; *b.i.d.*, twice daily; *c.i.v.*, continuous intravenous infusion; *CML*, chronic myeloid leukemia; *CR*, complete remission; *DAC*, decitabine; *i.v.*, intravenous; *MDS*, myelodysplastic syndrome; *NA*, not indicated; *t.i.d.*, three times daily.

DEBUSSCHER et al. (1990) treated 26 patients (median age 55 years) with refractory acute leukemia and chronic myeloid leukemia (CML) in blast crisis with 300–500mg/m$^2$/day of decitabine infused over 12–24 h, with treatment duration extending from 2 days to 5 days. One patient with AML achieved a complete hematologic and cytogenetic remission and received subsequent consolidation by autologous bone marrow transplantation. Five additional patients achieved a PR. The major toxicity was myelosuppression, with severe prolonged neutropenia and thrombocytopenia in several patients. Episodes of cytopenia and infection were clearly dose-dependent. The non-hematologic toxicity pattern was similar to that observed by Momparler et al., except that mild hepatic toxicity also occurred.

Pinto and coworkers conducted a phase-I/II study with decitabine as a single agent; the study included *untreated* AML patients who were not eligible for conventional induction treatment due to poor performance status (ZAGONEL et al. 1990, PETTI et al. 1993). As reported in a preliminary analysis, two complete and three partial responses were obtained among 22 patients [12 with de novo AML, eight with AML secondary to myelodysplastic syndrome (MDS) and two with blast crisis of CML], resulting in a 24% objective response rate. The median response duration was 14 weeks (range 2–32 weeks). The median survival of the responding patients was 26 weeks.

Phase-II studies in relapsed and resistant leukemia (mostly AML) were also performed with decitabine as a single agent (250–500mg/m$^2$ twice daily for 3–6 days) or in combination with either amsacrine (120mg/m$^2$ for 2 days) or idarubicin (12mg/m$^2$ for 3 days). RICHEL et al. (1991) reported one PR in five patients treated with decitabine alone, but eight CRs in 11 patients treated with the decitabine/amsacrine combination. Following these very encouraging results, two additional studies examining this combination were performed in Europe (WILLEMZE et al. 1993, 1997). The larger of these two studies was conducted by the European Organization for Research and Treatment of Cancer (EORTC) Leukemia Cooperative Group and was designed as a randomized phase-II trial to compare the effect of additional amsacrine to that of idarubicin. A total of 96 patients were treated with comparable schedules in these three studies, and an overall complete-response rate of 45% was obtained. In the randomized phase-II trial, 27% complete responses were observed in the decitabine/amsacrine combination versus 45% in the decitabine/idarubicin combination. This difference was, however, not statistically significant. Both normal karyotype and an interval of more than 1 year from initial diagnosis to the start of study treatment had a strong effect upon the complete response rate. Six of eight patients with normal karyotype and an interval of more than 12 months since diagnosis achieved a complete response, compared with one of eight patients with abnormal karyotype and an interval of less than 12 months since diagnosis (WILLEMZE et al. 1997).

The major toxicities observed in this randomized study were hematologic and gastrointestinal and were most pronounced in patients treated with idarubicin. The median remission duration was relatively long for a second-line treatment, but the majority of patients relapsed within 12 months after reaching CR. The median survival after CR was 15 months, and the authors concluded that decitabine is a useful antileukemic agent in the combination schedule and dosage used in this patient population, with efficacy and toxicity comparable to those of high-dose Ara-C.

SCHWARTSMANN et al. (1997) recently reported preliminary results of an ongoing phase-II study using decitabine ($90mg/m^2$/day for 5 days) in combination with daunorubicin ($50mg/m^2$/day over 3 days). All six evaluable patients with untreated AML (five with normal karyotype and one with inversion 16) achieved CR, while one patient with AML died of cerebral hemorrhage prior to evaluation. At the time of reporting, duration of consolidated remissions was from 5 months to 24 months (median: 11.5 months).

# 4 Antileukemic Activity of Azacitidine and Decitabine in Myeloid Blast Crisis of CML

In the United States, two groups have conducted studies in CML with azacitidine (in combination with etoposide) and decitabine (as single agents), respectively. Prognosis is very poor once CML accelerates and progresses from the chronic phase to blast crisis (mostly with myeloid phenotype, thus resembling AML, less frequently as lymphoid blast crisis, which can often be reverted to the second chronic phase with ALL-type treatment). Once myeloid blast crisis develops, however, remission rates with standard AML-induction regimens are below 20% (KANTARJIAN et al. 1993).

SCHIFFER et al. (1982) treated 27 patients with CML in myeloid blast crisis with azacitidine ($750mg/m^2$ total dose over 5 days, divided into 15 doses) in combination with etoposide ($75mg/m^2$/day for 5 days). The objective response rate was 58%, with one complete and 15 partial responses. Overall median survival was 23 weeks (33 weeks for responders, 10 weeks for non-responders). The major toxicities were nausea and vomiting, muscle aches and, less frequently, mucositis. Due to a decrease in toxicity with continued treatment, many patients were subsequently treated on an outpatient basis.

At the M.D. Anderson Cancer Center, decitabine ($500-1000mg/m^2$ administered over 5 days) was used in the treatment of 31 patients with CML in myeloid blast crisis (SACCHI et al. 1998) and 17 patients with CML in the accelerated phase (KANTARJIAN et al. 1997a,b). Objective responses were observed in 26% of patients in blast crisis, with a median survival of 29 weeks. One of the patients with a complete response had suppression of the Philadelphia chromosome (Ph) to 25% of metaphases. Of 17 patients with the accelerated phase of CML, nine (53%) responded to high-dose decitabine, with six patients achieving a second chronic phase of CML, and Ph suppression in two of them. Prolonged myelosuppression was the major side effect, but severe non-hematological toxicity was not observed. During these studies, the initial dose of $1000mg/m^2$ was, therefore, subsequently lowered to $750mg/m^2$ and $500mg/m^2$ in order to ameliorate the prolonged myelosuppression.

# 5 Decitabine as a Preparative Agent Prior to Allogeneic Peripheral-Blood Progenitor Cell Transplantation in Patients with Acute and Chronic Leukemias

Giralt and colleagues at the M.D. Anderson Cancer Center conducted two studies to determine the safety and efficacy of decitabine, as a single agent or as part of a combination preparative regimen, prior to allogeneic peripheral blood progenitor cell (PBPC) transplantation. The phase-I/II-study protocol of single-agent decitabine prior to allogeneic PBPC re-transfusion included patients with early relapse after a first allogeneic bone marrow transplantation for AML (eight patients), CML (two patients) and ALL (one patient), respectively (GIRALT et al. 1997, 1998a). Median time to relapse from initial transplantation was 6 months (range 2–31 months). Only one patient was in remission at the time of decitabine administration. Seven patients received a total dose of $1000mg/m^2$ decitabine (administered over 5 days); the remaining patients received $1250mg/m^2$. In the first three patients, donor cell transfusion was performed 2 days after the end of decitabine infusion. Seven of 11 patients (64%) achieved CR. Median time to neutrophil recovery was 13 days from donor cell transfusion (range 10–30 days). Two of the first three patients treated required additional donor cell transfusions at day 21 because of lack of neutrophil recovery. In subsequent patients, donor stem cells were given 5 days after the last decitabine infusion, and no delayed engraftment has been observed in the responding patients. No grade-3 or -4 toxicities were observed. Six patients developed grade-2 hepatic toxicity, one in association with graft-versus-host disease.

Six of seven patients achieving CR have relapsed, and four of them failed to respond to subsequent salvage treatment. Median survival for all patients was 12 weeks (range 2–97 weeks). In patients achieving a complete response, the median disease-free survival was 9 weeks (range 4–52 weeks). Thus, salvage therapy with decitabine followed by allogeneic progenitor cell support is feasible, well tolerated and induces CRs in the majority of patients. Combinations with other treatment modalities, such as donor-lymphocyte infusions or cytokine administrations, are necessary to prolong these remissions.

The other phase-I/II-study protocol includes leukemia patients prior to *first* allogeneic transplantation. The conditioning regimen contains decitabine (400–$600mg/m^2$) over 2 days followed by busulfan (1mg/kg orally four times daily over 4 days), and cyclophosphamide (100mg/kg) on two subsequent days. In four patients transplanted according to this protocol (one with AML, three with CML in the accelerated phase), the median time from leukemia diagnosis to transplantation was 5 months (range 4–25 months). The median age was 42 years (range 32–53 years). Two patients engrafted on days 23 and 25, respectively. Two patients needed additional donor stem cell transfusions at days 21 and 28 because of lack of neutrophil recovery, and subsequently recovered on days 31 and 37. Two patients achieved complete hematologic and cytogenetic remission. One of them had extramedullary relapse 19 weeks post-transplantation and received radiotherapy; the other was in remission 24 weeks after transplant. One patient with CML did not revert to a Ph-negative bone marrow; the fourth patient died from progressive *Pseudomonas* cellulitis 8 weeks after transplantation.

# 6 Results of Phase-I/II Clinical Trials of Azacitidine and Decitabine in Patients with Solid Tumors

Antitumor activity of both azacitidine and decitabine has also been explored in phase-I/II trials of patients with previously treated or metastasized solid tumors. As with AML and MDS, decitabine trials in solid tumors were performed predominantly in Europe (most of them conducted by the EORTC) and Canada, while azacitidine trials in these entities were performed in the United States.

In 1987, GLOVER et al. reviewed the results of seven trials, most of them broad, phase-II trials, performed with azacitidine in patients with solid tumors (including colorectal, breast, lung, melanoma, head and neck, testicular and renal cell carcinomas in addition to sarcomas). Among 505 patients treated, 29 PRs (5.7%) and three minor responses were observed. Therefore, these studies did not reveal adequate activity to justify further investigation at that time.

In the first phase-I study of decitabine, RIVARD and coworkers (1981) also included three children with metastasized solid tumors. Patients received decitabine by continuous infusion, but even at doses achieving antileukemic effects, only very limited antitumor effects were seen. A phase-I study of 21 adult patients with advanced solid tumors was performed by Pinedo and collagues using three daily 1-h infusions of decitabine repeated every 3–6 weeks, with doses ranging from $50\,mg/m^2$ to $300\,mg/m^2$ per course (GROENINGEN et al. 1986). The dose-limiting toxicity was myelosuppression, with a remarkably delayed leukopenia between days 22 and 23, whereas the platelet nadirs occurred between days 14 and 22. Other, less frequent toxicities possibly related to the drug were: a mild and reversible decrease of renal function in three patients, and mild, transient nausea and vomiting at different dose levels, including low ones.

A partial response at the highest dose level was noted in one patient with metastasized, undifferentiated carcinoma of the ethmoidal sinus, with complete regression of the local recurrence and marked decrease in the size of a single abdominal lymph node. No disease recurrence was observed in a 15-month follow-up after resection of the metastasis and continuation of decitabine treatment. The remaining 20 patients had either short-lasting stabilization or progression of disease.

In another phase-I study (using an i.v. bolus schedule of decitabine for the first 17 patients, a 4-h infusion in the remaining 15 patients), 32 patients with various advanced solid tumors were treated with dose levels ranging from $0.35\,mg/m^2/day$ (bolus infusion) to $22.5\,mg/m^2/day$ (4-h infusion) for 5 days, repeated every

4 weeks. In this trial, delayed leukopenias and thrombocytopenia were also major toxicities. No objective responses were seen and, in four patients with adenocarcinoma of the lung, colon and unknown primary, stable disease was noted (median: 4 months; range 2–5 months).

Using three daily 1-h infusions of $75mg/m^2$ repeated every 5 weeks, the EORTC Early Clinical Trials Cooperative Group conducted a total of seven phase-II trials with 153 patients with solid tumors. Tumor types included malignant melanoma (20), head and neck cancer (29), colorectal carcinoma (43), testicular cancer (15), renal cell carcinoma (16), non-small-cell lung cancer (8), ovarian cancer (27) and cervical cancer (17). Evaluable responses were seen in 133 patients. Of these, only two patients showed a PR: one of the 17 patients with malignant melanoma, and one of the five patients with non-small-cell lung cancer (DODION et al. 1990). The same schedule was also used in a phase-II study in cervical cancer by VERMORKEN et al. (1991). In this study, no responses were seen among 17 patients.

Two recent studies have revisited the concept of activity of DNA methylation inhibitors in solid tumors. MOMPARLER et al. (1997) performed a phase-I/II study of decitabine in patients with metastatic non-small-cell lung cancer. Fifteen patients were treated with a single 8-h i.v. infusion of $200–660mg/m^2$ of decitabine. The major side effect was hematopoietic toxicity, necessitating a 5- to 6-week recovery period before the next treatment course. Steady-state plasma concentrations of decitabine were measured in some of the patients and were in the same range as those resulting in in vitro demethylation of a gene hypermethylated in lung cancer cell lines. The median survival of these patients was 6.7 months, and three patients survived beyond 15 months, suggesting some clinical activity of relatively high doses of decitabine with an 8-h infusion schedule against metastatic lung cancer.

The second study was performed in 14 patients with progressive metastastic prostate cancer that was refractory to standard treatment (THIBAULT et al. 1998). The 1-h infusion schedule of $75mg/m^2$ three times daily, which had also been used in the EORTC studies in solid tumors, was applied. Treatment courses were repeated every 5–8 weeks to allow full recovery from myelotoxicity. Stable disease with time to progression of over 10 weeks was noted in two of the 12 patients with evaluable responses, both of them of African descent.

# 7 Results of Phase-II and -III Trials of Azacitidine in Patients with High-Risk Myeloclysplastic Syndrome (MDS)

MDS are a heterogeneous group of disorders originating from an early hematopoietic progenitor cell (KOEFFLER 1996). These disorders occur most frequently in elderly patients. The incidence in patients 70 years and older is at least $20/100,000$ patients/year (AUL et al. 1992). Clinical hallmarks of MDS are cytopenias, variable expansion of preleukemic, and clonal myeloblasts in the bone marrow and peripheral blood, with a variably increased risk of progression to AML. Thus, most patients eventually succumb to infections, bleeding complications, sequelae of iron overload, or secondary AML. Most MDS patients, due to their age and co-morbidity, are not eligible for allogeneic hematopoietic stem cell transplantation, the only established curative option. Thus, supportive care is the only generally accepted treatment standard for these patients. The effect of available recombinant lineage-specific hematopoietic growth factors (HGFs) is limited to improvement of single lineages and has not resulted in improved survival (KOEFFLER 1996), and no benefit has been demonstrated for differentiation inducers, such as retinoids and vitamin D3, as single agents.

The disturbed maturation of the morphologically dysplastic hematopoietic cells in MDS is thought to reflect a partial block in their differentiation, resulting in accumulation of these precursors in the bone marrow in spite of low peripheral blood counts. This "block in maturation", with proliferation of preleukemic myeloblasts, provided a rationale for clinical trials of DNA methylation inhibitors in MDS.

Azacitidine has shown activity in about 50% of MDS patients in several uncontrolled studies published since 1984, with a total of 106 patients in two phase-II trials initiated by the Cancer and Leukemia Group B (CALGB) in 1984 (trial 8421) and 1989 (trial 8921), respectively (Table 2). The CALGB 8421 study explored the activity of $75mg/m^2$ azacitidine by continuous infusion for 7 days (SILVERMAN et al. 1993). Treatment was repeated every 28 days, with a minimum of four courses. Patients achieving CR continued treatment for three additional courses, whereas patients reaching PR or hematological improvement continued treatment until relapse or disease progression. Of 43 patients, 21 (49%) responded to this treatment (Table 2). Trilineage responses, i.e., partial or complete correction of all three deficient blood cell lineages and the blast excess, occurred in 37% of patients. Interestingly, the median time to response was 3.8 treatment courses (range 2–11 courses), indicating that the mechanism of action of azacitidine probably necessitates repeated application over time to achieve improvement of hematological parameters. This may explain the lower response rates in other trials when less then three treatment courses were applied. The median response duration was 14.7 months, and the median overall survival was 13.3 months. The most frequent side effect was nausea and/or vomiting (63%), followed by diarrhea (30%). Other side effects were less frequent (transaminase elevations and confusion in 16% and 6% of patients, respectively).

The same dose schedule but with administration by daily subcutaneous bolus injection of azacitidine, thus allowing for outpatient treatment, was used in the CALGB 8921 trial (SILVERMAN et al. 1994; Table 2). Response rates in this trial were comparable to those of the previous trial, except for a somewhat lower rate of trilineage responses (27% vs 37%). RUGO et al. (1999) also treated 92 patients, most of them with high-risk MDS or secondary AML, with the same outpatient schedule of $75mg/m^2$/day given subcutaneously for 7 days (repeated every 28 days for six cycles) between 1991 and 1998. In a retrospective analysis of this compassionate-use program, they reported 61% responses (13% and 19% CR and PR, respectively). Two patients with complete hematologic response also had a cytogenetic remission.

**Table 2.** Clinical trials of azacitidine and decitabine in patients with high-risk myelodysplastic syndromes.

| Authors | n | Age[a] (range) (years) | RAEB/RAEB-t/ CMML (%) | Drug schedule | Percentage responses (% CR/PR/HI) | Number of courses to best response[a] (range) | Response duration (months) | Overall survival (months) |
|---|---|---|---|---|---|---|---|---|
| CHIMBATAR et al. 1991 | 15 | 64 (31–76) | 53 | azaC 10–35mg/m²/ day × 14 (c.i.v.) | 23 (0/0/23) | 1.5 | NA | NA |
| SILVERMAN et al. 1993 | 43 | 65 (NA) | 100 | azaC 75mg/m²/ day × 7 (c.i.v.) | 49 (12/25/12) | 3.8 | 14.7 | 13.3 |
| SHADDUCK et al. 1999 | 24 | 68 (36–82) | 65 | azaC 75mg/m²/ day × 7 (s.c.) | 45 (10/0/35) | NA | NA | NA |
| SILVERMAN et al. 1994 | 68 | 66 (23–82) | 100 | azaC 75mg/m²/ day × 7 (s.c.) | 53 (12/15/27) | 4.5 | 17.3 | NA |
| SILVERMAN et al. 1998 | 99[b] | 68 (NA) | 63 | azaC 75mg/m²/ day × 7 (s.c.) | 63 (6/10/47) | NA | 14.5 | 18[b] |
| RUGO et al. 1999 | 92[c] | 27–91 (NA) | 60 | azaC 75mg/m²/ day × 7 (s.c.) | 61 (13/29/14) | NA | (2–30+) | NA |
| ZAGONEL et al. 1993 | 10 | 68 (60–78) | 100 | DAC 45mg/m²/ day × 3 (i.v.); DAC 50mg/m²/ day × 3 (c.i.v.) | 50 (40/10/0) | 2 | 11 | NA |
| WIJERMANS et al. 1997 | 29 | 72 (58–82) | 72 | DAC 40–75mg/m²/ day × 3 (c.i.v.) | 54 (28/18/7) | 1.8 | 7.3 | 10.5 |
| WIJERMANS et al. 1999 | 66 | 68 (38–84) | 88 | DAC 45mg/m²/ day × 3 (i.v.) | 49 (20/5/24) | 3.2 | 7.3 | 15 |

*azaC*, azacitidine; *c.i.v.*, continuous intravenous infusion; *CMML*, chronic myelomonocytic leukemia; *CR*, complete remission; *DAC*, decitabine; *HI*, hematologic improvement; *i.v.*, intravenous; *NA*, not given; *PR*, partial remission; *RAEB*, refractory anemia with excess blasts; *RAEB-t*, refractory anemia with excess blasts in transformation; *s.c.*, subcutaneous injection.

[a] Median.

[b] Patients randomized for treatment arm.

[c] Retrospective analysis of compassionate-use program. Overall survival was from start of treatment.

The major toxicity in this retrospective analysis was mild to moderate nausea and vomiting. To address the question of whether treatment with azacitidine may change the natural history of MDS, a randomized phase-III study was initiated by CALGB in 1994 using the identical dose and subcutaneous route. In this study, response rate, survival and time to progression to AML in treated patients were compared with those of a patient group assigned to a four-month observation period (SILVERMAN et al. 1998). Study design allowed for patients in the observation arm to cross over to azacitidine treatment after 4 months (and in some instances 2 months) of observation in case of disease progression. The response rate in the group of patients who were treated immediately was 63%. The probability of transformation to AML was 11% in azacitidine-treated patients, compared with 31% in patients on observation, which was statistically significant ($P = 0.003$).

Quality-of-life analyses revealed significant improvement of well-being in the treatment group. The median survival of treated patients was 18 months, compared with 14 months in those on observation (not statistically significant). However, 58% of patients crossed over from observation to azacitidine treatment, making comparison of survival in both groups difficult. In summary, this trial implicated that the natural history of MDS may be changed by a non-intensive drug treatment, which previous large, comparative trials testing growth factors (reviewed by KOEFFLER 1996) and low-dose Ara-C (MILLER et al. 1992), respectively, had failed to demonstrate.

# 8 Results of Low-Dose Decitabine Trials in Patients with High-Risk MDS

Clinical trials of decitabine in MDS have thus far been performed solely in Europe. Following the encouraging results of an intermittent infusion schedule of decitabine ($15-60mg/m^2 \times 9$ infusions over 72 h, thus allowing for alternate 4-h treatment and resting periods) for treating patients with de-novo AML and AML from MDS (see above), the Aviano group reported results of a small series of ten consecutive patients with high-risk MDS treated with two comparable, low-dose, 3-day schedules (ZAGONEL et al. 1993). Six patients received $45mg/m^2/day$, divided into three infusions; four patients received $50mg/m^2$ day by continuous infusion. The authors noted a 50% response rate, with 40% complete responses (Table 2). The starting dose of $50mg/m^2/day$ by continuous infusion (with dose escalations to $65mg/m^2/day$ and $75mg/m^2/day$ in non-responding patients) for 3 days was also used by Wijermans and colleagues (1997). They observed an overall response rate of 54%, with 46% trilineage responses; however, serious myelotoxicity occurring at the higher dose levels necessitated dose reduction to $40mg/m^2/day$. Thus, cytopenia was the major side effect in this trial, with nausea occurring in only 10% of patients.

In a phase-II multicenter trial also chaired by Dr. Wijermans, a total of 66 previously untreated patients with high-risk MDS were treated with a total dose of

135mg/m$^2$ decitabine (administered as three daily 15mg/m$^2$ infusions given over 4 h and for a total of 3 days) repeated every 6 weeks. Primary aims of this study were to determine response rate and toxicity; secondary aims were response duration, survival after starting therapy and overall survival.

Sixty-six patients (median age 68 years) from seven centers in the Netherlands, Belgium and Germany were to receive a minimum of two courses (Table 2). In case of CR at that time, this was consolidated with two additional cycles. In case of stable disease, hematological improvement or partial response, a maximum of six cycles was administered. The overall response rate in this study was 49%, with an actuarial median survival of 22 months from diagnosis and 15 months from the start of therapy. The actuarial median response duration was 31 weeks the median progression-free survival 25 weeks. Myelosuppression was the major toxicity, with a treatment-related mortality of 7% (due to pancytopenia and infection). Interesting responses were observed in the megakaryopoiesis, with rises in platelet counts after only one cycle of decitabine therapy in the majority of the responding patients. Subgroup analysis revealed that the highest response rates were in the "high-risk" patient group, according to the International Prognostic Scoring System (GREENBERG et al. 1997). 16 of 25 patients (64%) in this group responded compared to 12 of 25 (48%) in the "intermediate-2" risk group, and four of 16 (25%) in the "intermediate-1" group. This difference was statistically significant ($P = 0.014$ by the Chi-Squared test). No significant difference in response duration was noted for the different patient groups (WIJERMANS et al. 1999).

Cytogenetic analyses were performed in both of the above studies before, during and after treatment with decitabine in order to determine the incidence and duration of cytogenetic responses (LÜBBERT et al. 1999). Since the abnormal clone in MDS can coexist with bone marrow cells of normal karyotype (representing residual normal hematopoiesis), efficient removal of the defective cell clone and suppression of pre-leukemic blast expansion would be expected to shift the ratio of cells with abnormal versus normal karyograms. Of 68 patients karyotyped in both studies prior to treatment, 38 (56%) had abnormal karyograms and were thus informative for the analysis. Following treatment with 2–6 courses of decitabine, major cytogenetic responses were observed in 11 patients (29%). Suppression of the abnormal clone may have occurred by an inhibitory, cytotoxic, or differentiation-inducing effect of decitabine, or a combination of these. In this regard, it was interesting to note that the average number of treatment cycles necessary for the best cytogenetic response was 3.3 courses (range 2–6 courses), indicating that repeated courses over time may be necessary to achieve the best results in a majority of patients. Sequential analyses performed on two patients using both cytogenetics and fluorescence in situ hybridization also disclosed a continuous and linear reduction of the abnormal clone over the entire treatment period of six courses of decitabine (i.e., approximately 9 months), with the best response in both patients being achieved after the final treatment course.

Ten of the 11 patients with major cytogenetic responses relapsed with the initially detected abnormal clone. Median time to relapse was 8 months. This probably indicates incomplete elimination of the abnormal clone by the treatment. However, rapid acquisition of additional chromosomal abnormalities following treatment with decitabine appears to be an infrequent event. This is relevant, considering that both azacitidine and decitabine have been proposed to have mutagenic potential and, at high concentrations in vitro, have been shown to induce chromosomal instability. This would be a caveat when considering treatment of younger patients with good prognosis and non-neoplastic disorders, such as mild hemoglobinopathies (see below). However, given the small number of patients in this study and the short observation period in patients who have, due to nature of their disease, a limited prognosis (5–15 months median survival in untreated patients), longer follow-ups during this study and subsequent, larger trials are necessary to further define the incidence and type of possible mutagenic action of these drugs.

Overall, azacitidine and decitabine have demonstrated significant activity in ameliorating or even temporarily correcting both the deficits in peripheral blood counts of all three lineages and the pre-leukemic blast excess in high-risk MDS. This combination of activities makes these drugs unique in a disease in which non-intensive treatment, especially for elderly patients with often significant co-morbidity, may be the optimal therapy. Apart from "AML-type", intensive induction treatment, no other single drug or drug combination has induced significant tri-lineage responses together with an antileukemic activity.

# 9  Clinical Effects of Azacitidine and Decitabine in Severe β-Thalassemia and Sickle Cell Anemia

VAN DER PLOEG and FLAVELL (1980) first demonstrated the presence of undermethylated regions upstream of the duplicated γ-globin (μ) gene locus (encompassing the $^G$μ and $^A$μ genes) in fetal liver cells which express these genes. Remethylation at the restriction sites analyzed was present in adult human bone marrow cells, in which the μ-globin genes are not expressed due to transcriptional shut-off. This discovery provided a rationale to explore the pharmacologic stimulation of fetal hemoglobin (HbF) production, through induction of γ-globin messenger RNA (mRNA) expression with demethylating agents, in patients with severe β-thalassemia. In an animal model of anemic baboons, short-term azacitidine administration resulted in a marked (albeit transient) enhancement of HbF synthesis (DeSimone et al. 1982). In 1982, Ley et al. first reported clinical results on a patient with severe β-thalassemia and iron overload due to polytransfusion. Azacitidine was administered by continuous infusion at 2mg/kg/day for 7 days (Table 3). Following infusion, a rise in hemoglobin (Hb) levels from 8g/dl to 10.8g/dl and a more than fourfold increase in reticulocytes were attained. Hb values remained above pretreatment levels for nearly 5 weeks without transfusion. Side effects included moderate nausea and vomiting on 2 days of treatment and mild leukopenia 3 weeks after the end of treatment. In vitro studies revealed a sevenfold induction of μ-globin mRNA, with a marked and equal increase of both $^G$μ- and $^A$μ-chain

**Table 3.** Clinical effects of azacitidine and decitabine in severe β-thalassemia and sickle cell anemia.

| Authors | n | Age (range) (years) | Hemoglobinopathy | Drug schedule | Percentage maximal HbF | Maximal increase in Hb (g/dl) | Demethylation |
|---|---|---|---|---|---|---|---|
| Ley et al. 1982 | 1 | 42 | β+ thalassemia | azaC 2mg/kg/day × 7 (c.i.v.) | 20.8 | 2.8 | $^G\mu$, $^A\mu$, ε |
| Charache et al. 1983 | 1 | 32 | Sickle cell anemia | azaC 30mg/m²/day × 3 (4 h i.v., s.c.) | 8.9 | 3.0 | Global, $^G\mu$, $^A\mu$, Y rpt |
| Ley et al. 1983 | 4 | 32 (22–58) | β+ thalassemia (2), sickle cell anemia (2) | azaC 2mg/kg/day × 7 (c.i.v.) | 21.4 (8.9–64) | 1.9 (1.2–2.9) | $^G\mu$, $^A\mu$ |
| Dover et al. 1985 | 4[a] | 29 (23–45) | Sickle cell anemia | azaC 2mg/kg/day × 15–35 (s.c.), azaC 0.2mg/kg/day (p.o.) + THU 200mg × 35 (p.o.) | 13.6 (9.6–17.5) | 3.0 (1.2–3.5) | ND |
| Humphries et al. 1985 | 7[b] | 34 (28–40) | Sickle cell anemia | azaC 2mg/kg/day × 7 (c.i.v.), azaC 1.5mg/kg/day × 5 (8 h i.v.) | ND | 1.4 (0.2–2.5) | $^G\mu$, $^A\mu$ |
| Dunbar et al. 1989 | 1 | 38 | β° thalassemia | azaC 2mg/kg/day × 5 (c.i.v.) | ND | 3.0 | ND |
| Lowry et al. 1993 | 3[c] | 32 (30–50) | β+ thalassemia (2); β° thalassemia (1) | azaC 1–2 mg/kg/day i.v. × 4 (c.i.v.) | 76 (52–100)[d] | 3.1 (2.9–4.4) | ND |
| Koshy et al. 1998 | 9 | NA | Sickle cell anemia | DAC 0.15mg/kg/day × 10 (i.v.) | 9.1 (NA) | NA | ND |

*azaC*, azacitidine; *β+ thalassemia*, β-thalassemia intermedia; *β° thalassemia*, β thalassemia major; *c.i.v.*, continuous intravenous infusion; *DAC*, decitabine; $^G\mu$, $^A\mu$, ε, globin genes; *Hb*, hemoglobin, *HbF*, hemoglobin F; *i.v.*, intravenous; *NA*, not indicated; *ND*, not determined; *p.o.*, per os (orally); *s.c.*, subcutaneous injection; *THU*, tetrahydrouridine; *Y rpt*, anonymous repetitive sequence derived from the Y chromosome. Global demethylation was assessed by restriction of genomic DNA with methylation-sensitive restriction enzyme and gel electrophoresis

[a] Patient A had previously been described (Charache et al. 1983).

[b] Five additional patients had previously been described.

[c] Patient 1 had previously been described (Ley et al. 1982).

[d] Measurements were made in the absence of transfusions and were not determined in patient 2.

synthesis. Transient hypomethylation of the γ-globin gene locus coincided with augmented HbF production (see below).

CHARACHE et al. (1983) treated a 32-year-old patient with sickle cell anemia and severe transfusion requirements with i.v. azacitidine infusions ($3 \times 30mg/m^2$ for 4 h at 8-h intervals). In this patient, a similar rise in HbF, reticulocytes and Hb levels was obtained and, upon decrease to baseline values, could be reproduced with repeated treatment. Since these pioneering studies, five other reports have described similar effects in six additional patients with severe β-thalassemia and 12 additional patients with sickle cell anemia (Table 3). Myelosuppression was usually mild during the first courses and only necessitated changes in the dose or schedule after repeated courses. In 1993, LOWREY and NIENHUIS reviewed the long-term responses and outcome of three thalassemia patients (including the patient first described in 1982) for whom continued transfusion therapy was no longer beneficial. All three patients became transfusion-independent (two of them for more than a year) with repeated courses of i.v. azacitidine. The first patient had a remarkable improvement in cardiac function following effective iron depletion (by chelating agents and phlebotomy) only after the addition of azacitidine. The authors concluded that the efficacy and feasibility of this treatment clearly justifies its use in carefully selected patients for whom no other treatment options are available. However, given the potential risks of carcinogenicity, which could not be addressed in the patient groups having received the drug (due to their poor prognosis), they cautioned against its use in thalassemia patients for whom other treatment options are available and active (LOWREY and NIENHUIS 1993).

Recently, KOSHY et al. (1998) explored the effects of decitabine in patients with sickle cell anemia unresponsive to hydroxyurea (HU). In a preliminary report, they noted a threefold increase in HbF in all nine patients treated with very low doses of i.v. decitabine (starting dose: 0.15mg/kg/day) for 5 days, repeated once during the course of 2 weeks. Maximum levels of HbF were reached within 4 weeks and were sustained for at least two more weeks. The only side effect noted was mild, reversible neutropenia. They concluded that, with the efficacy of this low-dose, short-term schedule in a patient group not responding to HU, further increases in HbF may be attained upon dose escalation.

# 10 Mechanisms of Action of DNA Methylation Inhibitors in Hemoglobinopathies, MDS and Leukemia

Studies addressing the mechanism of action of DNA-demethylating agents in the clinical setting were first performed in conjunction with treatment of individual β-thalassemia and sickle cell patients treated with azacitidine. CHARACHE et al. (1983) estimated global methylation changes in peripheral blood leukocytes and bone marrow mononuclear cells (MNC) before, during and after treatment using agarose-gel electrophoresis of DNAs cleaved with methylation-sensitive restriction enzymes. They noted a decrease in global methylation between 5 days and 7 days after treatment, which was reversed during longer observation periods. When probing specific genes for methylation changes, partial demethylation at two HpaII sites situated 107 bp 5′ of the $^G\mu$ and $^A\mu$ genes was reproducibly detected as early as 2 days after beginning treatment. It reversed within weeks after cessation of treatment (LEY et al. 1982, 1983; CHARACHE et al. 1983; HUMPHRIES et al. 1985). Demethylation was paralleled or followed by induction of μ-globulin mRNA in bone-marrow MNC, as quantified by S1 mapping (HUMPHRIES et al. 1985). In these studies, other genes were also probed to determine the specificity of this observation, and demethylation of an anonymous repetitive sequence derived from the Y chromosome was noted (Table 3). The γ gene became equally demethylated (without, however, significant induction of its mRNA; LEY et al. 1982).

Given the heterogeneity of the erythroid progenitor population in the bone marrow, it was unclear whether induction of HbF and the observed demethylation were due to (1) perturbations of the erythroid progenitor pools due to a nonspecific cytotoxic effect of azacitidine (TORREALBA-DE RON et al. 1984), (2) a specific direct effect upon the erythroid progenitors stimulated to synthesize globin chains, or even (3) gene-specific hypomethylation of individual alleles. In an elegant study by HUMPHRIES et al. (1985), this question was addressed by employing cell-cloning techniques in soft agar. By picking individual colonies from erythroid precursors at early time points during treatment, the authors could, in fact, demonstrate increased HbF synthesis in the absence of suppression of this erythroid population via cytotoxicity. These effects were also observed in normal human bone marrow treated with azacitidine in vitro. These results, together with an early increase in HbF-containing reticulocytes in the peripheral blood and early induction of globin mRNA led the authors to conclude that a direct effect of the drug upon the erythroid progenitor cells was very likely. Since PCR and high-speed cell-sorting techniques were not available at that time, the methylation status at the single-cell level could not be studied.

Several questions as to the mechanism of action of DNA demethylating agents were also addressed with regard to the effects observed in patients with MDS and leukemia. These can be summarized as follows:

1. What are the underlying mechanisms resulting in improved erythropoesis, thrombopoiesis and granulopoiesis in patients with MDS treated with these drugs?
2. What are the mechanisms of blast suppression in MDS with pre-leukemic blast excess and in leukemias, and is differentiation involved?
3. Which are the target genes mediating these diverse effects that are possibly mediated by demethylation?

SILVERMAN et al. (1993) examined bone marrow cells from three patients with MDS responding to azacitidine with a rise in Hb for erythroid-progenitor growth patterns before and during treatment with azacitidine. The BFU-e (burst-forming unit erythroid, i.e., early erythroid-progenitor cells) and CFU-e (i.e., the more mature colony-forming unit erythroid) were abnormally low or absent. Continued treatment with azacitidine was associated with an increase in BFU-e and normalization in CFU-e numbers preceding the improvement in Hb levels, resulting in transfusion independence in two patients. It could be argued that temporary normalization of erythroid-progenitor growth and Hb levels could be due to expansion of normal hematopoiesis. However, persistent dysplasia also involving the erythroid compartment in two of three cases suggests correction of the previous, ineffective erythropoiesis at the level of the target cell. This could occur either by direct action of azacitidine on the dysplastic precursors or via a humoral mediator acting on these cells, e.g., secreted by bone marrow stroma.

The Aviano group tested the hypothesis of in vivo differentiation through decitabine using morphology, immunocytochemistry and molecular techniques on leukemia cells. Based upon the initial observation of in vitro differentiation [by low concentrations of decitabine (1.0μmol/l) pulsed for 5–8 days, accompanied by upregulation of major histocompatibility (MHC) class-I molecules (PINTO et al. 1984)] of primary monoblastic and myeloblastic leukemia cells from four of five AML patients, similar morphologic and immunophenotypic studies were performed in patients with MDS, AML and CML treated with decitabine (PINTO et al. 1990; ZAGONEL et al. 1990; PETTI et al. 1993). Morphological changes of bone marrow leukemic blasts and surface marker-profile alterations suggestive of some differentiation were seen in the majority of patients. Specifically, leukemic cells in some AML patients had an increase in cytoplasmic granules, nuclear maturation and/or segmentation. In a single patient, a gradual increase in myeloperoxidase activity was noted in the absence of other signs of maturation. These morphological changes were restricted to AML patients showing a clinical response to the treatment (PINTO et al. 1990).

Expression of early (CD34, CD33), intermediate (CD13) and late (CD11b, CD11c, CD14, CD16) markers of myeloid cell maturation was measured on AML blasts using fluorescence-activated-cell-sorting analysis (PINTO and ZAGONEL 1993). A rather heterogeneous pattern of modulation of these molecules was observed following decitabine treatment and was not restricted to responding patients. Downregulation of early and intermediate markers (CD13 and CD33) with concomitant upregulation of late markers (CD16, CD11c, CD14) was repeatedly observed. In several patients with AML and aberrant expression of the lymphoid marker CD25 on the blast cells, this molecule was downregulated following decitabine treatment. Finally, in a single case, the early marker CD34 persisted in the presence of continuous downregulation of CD33 and CD13 and upregulation of the late marker CD14, giving rise to a population with "asynchronous" immunophenotype. MHC class-I molecules were also frequently found to be upregulated in clinically responding patients in these studies.

RICHEL and colleagues (1991) made similar observations in six patients with acute leukemias treated with decitabine and amsacrine. In these patients, leukemic infiltration was unchanged at day 7 of treatment, thus allowing for a quantitative analysis of marker expression. In three of six patients, early markers (CD33 and/or CD34) were downregulated from a mean fluorescence intensity of 66% (range, 25–81%) to 1–2%. In one of the patients with ALL, a strong but transient downregulation of B cell markers CD19, CD20, and CD22 was also observed. Thus the leukemic immunophenotype may be altered by decitabine treatment. However, changes were quite heterogeneous and not strictly associated with hematologic response. Although they are suggestive of differentiation in some leukemias, this needs to be confirmed by studies showing that a mature progeny (peripheral blood granulocytes) with dysplastic morphology and aberrant immunophenotype is bearing the clonal marker.

An obvious question regarding the molecular mechanism of action of DNA methylation inhibitors is whether the serum levels achieved with these drugs in the clinical protocols used are sufficient to induce global cytosine demethylation comparable to those that are effective in vitro. In three clinical studies of acute leukemias treated with azacitidine (AVRAMIS et al. 1989) or decitabine (MOM-PARLER et al. 1986; RICHEL et al. 1991), leukemic blasts of a total of 20 patients were subjected to analysis of inhibition of total cytosine methylation before and after treatment. Schedules were of 7-day, 40-h and 5-day duration, respectively. Ratios of 5-methyl deoxycytidine and deoxycytidine were quantified by high-pressure liquid chromatography after complete digestion of total DNA and, in one study, also using in vivo labeling with [6-$^3$H] deoxycytidine. Bone marrow or blood samples obtained both before and at the end of treatment were analyzed. A significant decrease in methylation after treatment was observed in all 20 patients (mean of methylation inhibition: 55.6%, 76% and 29%). In the study by RICHEL et al. (1991) plasma levels of were also determined; these were dependent on the infusion rate and ranged from 0.6μM to 5.0μM. However, in all three studies, the inhibition of methylation observed did not correlate with the clinical outcome of the patients treated.

Regarding the specific target genes possibly modulated by these drugs via demethylation, numerous in vitro studies have clearly shown that either compound can result in demethylation of hypermethylated and silenced genes, with subsequent mRNA re-expression. This includes the p16 and p15 tumor-suppressor genes, which are frequently hypermethylated in a variety of epithelial cancers (p16) and myeloid malignancies (p15). Using the T24 bladder cancer cell line, in which the p16 gene is hypermethylated and transcriptionally silenced, BENDER et al. (1998) have elegantly demonstrated a time-dependent, heritable effect of decitabine upon demethylation and re-expression of p16 mRNA. Cell-cycle studies revealed a significant shift of the cells from the G2/M to the G0/1 phase of the cell division cycle following p16 re-expression.

p15, but not p16, is frequently hypermethylated in both AML (HERMAN et al. 1997; AGGERHOLM et al. 1999) and MDS (UCHIDA et al. 1997; QUESNEL et al. 1998). Its silencing is associated with lack of expression and a shift of cells from the G1/S to the G2/M phase of the cell-division cycle. Therefore, one could hypothesize that p15 may be subject to demethylation by DNA methylation inhibitors in MDS patients with bone-marrow blast excess successfully treated with these drugs. Prior to decitabine treatment, 22 patients with high-risk MDS were examined for p15 methylation. In 13 (59%), hypermethylation was detected. Sixteen patients could be analyzed sequentially after one or more treatment courses with decitabine. In seven of ten patients with p15 hypermethylation prior to treatment, demethylation by

25% or more was detected using a quantitative, methylation-sensitive, single-nucleotide primer-extension assay (DASKALAKIS et al. 1998). Interestingly, all of these patients also had a clinical response, including blast suppression in patients with an initial excess of bone marrow blasts. Among the six patients where no hypermethylation was detectable prior to treatment, sequential analysis did not reveal a change in methylation status; however, four of them also responded to the treatment. Thus, hypermethylated p15 gene may be one possible target of the demethylating activity of decitabine, but responses can also occur in the absence of p15 hypermethylation.

Unravelling the specific cellular and molecular targets of DNA methylation inhibitors in MDS will be a major task, considering the heterogeneity of this disease. However, the kinetics of hematologic improvement may be summarized as follows: (1) an "early" thrombocyte increment following the initial 1–2 courses of treatment, with (2) subsequent improvement or normalization of granulocyte counts with continued treatment, paralleled or followed by (3) a gradual improvement of Hb level. Reversion of blast excess is also gradual, with complete normalization (less than 5% bone-marrow blasts, no peripheral-blood blasts) usually occurring after more than two courses of treatment.

DNA methylation inhibitors, when compared with Ara-C, which does not decrease DNA methylation, have a molecular structure and enzymatic inactivation similar to this drug. However, there are clear differences in clinical responses, suggesting that their mechanisms are distinct. Specifically, the rapid and strong blast reduction induced by Ara-C frequently does not occur with azacitidine or decitabine treatment. Low-dose Ara-C produces significant reduction of bone marrow cellularity, which reflects its myelotoxic activity; this is associated with a clinical response (MILLER et al. 1992). In contrast, responses to azacitidine or decitabine have been described in the absence of significant reduction of bone marrow hypoplasia, even during phases of peripheral blood cytopenia (PINTO et al. 1990; SILVERMAN et al. 1993). Responses in MDS occur late during continued treatment over time and have also been observed several months after discontinuation of treatment (PINTO and ZAGONEL 1993). Indirect evidence stems from an algorithm for predicting response to low-dose Ara-C (HELLSTRÖM-LINDBERG et al. 1994), by which the lowest probability of response ( < 2%) is predicted in patients with (1) at least two chromosomal abnormalities, (2) thrombocytopenia and (3) hypercellular bone marrow. However, patients with all of these features often respond quite well to decitabine or azacitidine. In summary, DNA methylation inhibitors have clinical activity that clearly differs from that of low-dose Ara-C.

# 11 Summary and Future Directions

From results of clinical studies performed over more than 20 years with both azacitidine and decitabine in acute leukemias and MDS, one can conclude that both have comparable activity in these diseases. Relapsed and refractory AML and previously untreated high-risk MDS patients have been the most extensively studied subgroups with respect to drug schedule and effectivity. In relapsed/re-fractory AML (and CML in blast crisis), schedules with total doses ranging between 500mg/m$^2$ and 1500mg/m$^2$ with either drug are as effective (or are superior to) high-dose Ara-C. Lower dose schedules in the treatment of AML have been explored only in a limited number of studies, with inconclusive results regarding the best schedule and effectivity.

The pioneering studies of the Aviano group have demonstrated the effectivity of several low-dose schedules in high-risk MDS (which often precedes AML of the elderly, since these patients often present with a clinical or morphologically detectable myelodysplastic phase). The majority of these AML patients are not eligible for intensive induction–consolidation treatment, due to their age and co-morbidity. Therefore, it would be of great interest to systematically study lower dose, first-line schedules of decitabine or azacitidine in this patient group. Outpa-tient schedules using subcutaneous injection would of course be very useful in this regard. The initial, rapid blast lysis that is typically induced by Ara-C often does not occur with methylation inhibitors. Therefore, combinations with hydroxyurea or Ara-C would probably be necessary to control clinically relevant leukocytosis present at the start of treatment. Kinetics of blast removal in the MDS trials show that these drugs are most effective when given over a prolonged period of repeated courses, which might be considered in the design of such protocols. Once the best response is achieved, DNA methylation inhibitors, given at even lower doses, may also be useful agents in the maintenance of these responses.

The randomized phase-III study performed by the CALGB (SILVERMAN et al. 1998) has implicated azacitidine as a drug to alter the natural course of high-risk MDS. The very encouraging results of phase-II studies with decitabine also strongly urge for proof of its effectivity in a controlled study. Since about 50% of high-risk MDS patients do not respond to demethylating agents, rational drug combinations should be another step in further improving these results. Given the known myelotoxicity of these drugs in a disease presenting with cytopenias, clin-ically effective combinations with compounds that have little or no myelotoxicity are highly desirable. These may include HGFs and/or differentiating agents, such as all-*trans* retinoic acid which, as a single agent, probably has little activity in MDS, but may be more effective in the presence of decitabine due to upregulation of its receptor (CÔTÉ and MOMPARLER 1997). Since most MDS patients eventually relapse following treatment with azacitidine or decitabine, a prolongation of re-mission may possibly be achieved with a lower dose schedule as maintenance therapy.

Other future studies might define a possible role of even lower dose schedules (with less myelotoxicity) in low-risk MDS and in other disorders that are responsive to DNA methylation inhibitors. KOSHY et al. (1998) recently reported that dec-itabine, at starting doses of 1.5mg/kg per course (divided into ten doses of 0.15mg/kg administered over 14 days), augments HbF levels in sickle-cell anemia patients. Other recurrent effects seen at this very low dose were mild neutropenia and an increase in platelet count. The promising early results of this interesting study imply that this drug exerts its mechanism(s) even at a total dose that is ~50% of that used in high-risk MDS (notwithstanding different time schedules of administration). Further studies are necessary to define this activity in sickle cell patients that are refractory to HU with respect to duration of treatment, development of resistance, and potential carcinogenicity.

The ongoing studies by Giralt and coworkers on decitabine in the allogeneic transplantation setting show that it is feasible to use this drug in preparative regimens in leukemia and MDS patients. Since the relapse rate of AML and MDS patients in non-intensive preparative regimens is high, the use of this compound, which can upregulate MHC class-I molecules in residual malignant cells and, therefore, improve antileukemic effects of donor-lymphocyte infusion, should be further defined.

The phase-I/II studies of azacitidine and decitabine performed in the 1970s and 1980s, respectively, in patients with solid tumors have yielded disappointing results overall. However, with the knowledge derived from studies of single-agent DNA-methylation inhibitors in MDS and AML regarding effective drug schedules, the very limited non-hematologic toxicity and the necessity to administer these drugs over a prolonged period to achieve a progressive removal of malignant cells, it would be of interest to re-evaluate the activity of these drugs in solid tumors. The rationale for revisiting this issue could possibly be strengthened by recent investigations from several laboratories demonstrating hypermethylation and transcriptional silencing of tumor-suppressor genes (p16/INK4A, p15/INK4B, Rb, VHL) in different types of solid tumors. Results obtained on decreased methylation of p15 in mononuclear bone marrow cells from MDS treated with decitabine suggest hypermethylated genes as appropriate targets of DNA methylation inhibitors even at non-intensive dose schedules. Given their short plasma half-life, repeated administration of decitabine or azacitidine with prolonged infusion duration in solid tumors with known hypermethylation of p16, e.g., bladder cancer or non-small-cell lung cancer, might result in antitumor activity that is superior to the disappointing results obtained with 1-h infusion schedules.

The available data on the mechanism of action of these drugs strengthen the idea that it is different from that of agents that act primarily via their cytotoxic effects, such as low-dose Ara-C. In 1984, Momparler et al. described the effect of decitabine in leukemia as probably involving "... gene activation and induction of differentiation. One would not expect to observe an acute cell kill, but a disorganization of gene expression and a gradual decrease in cell number due to senescence." In fact, most investigators treating patients with MDS with these drugs have observed remissions obtained in the absence of true bone marrow aplasia and late remissions occurring months after stopping administration of these drugs. Since hypermethylation and silencing of tumor-suppressor genes involved in cell-cycle regulation is frequent in leukemia and MDS, demethylation and reactivation of such genes might, at least in part, explain these phenomena.

It is tempting to speculate what other groups of genes may be subject to demethylation in diseases that are responsive to DNA methylation inhibitors. Pinto has reported upregulation of granulocyte-colony-stimulating-factor receptor on bone marrow cells from a patient with MDS treated with decitabine (PINTO and ZAGONEL 1993), which would be an attractive, simple explanation for the observed improvement of granulocytopenia in responding patients. Similarly, improvement of anemia and rapid induction of thrombocytosis in this disease following treatment with DNA-methylation inhibitors could be speculated to be due to upregulation of lineage-specific receptor molecules. Clonality studies on granulocytes mobilized in responding MDS patients may clarify whether the activity of DNA methylation inhibitors is via differentiation induction. Finally, with further evidence that DNA demethylation induced by both drugs is linked to their clinical activities, combinations with other compounds inhibiting methylation but lacking myelotoxicity, such as antisense oligonucleotides inhibiting Dnmt1 (RAMCHANDANI et al. 1997), would be very interesting combinations in diseases where azacitidine and decitabine are active.

*Acknowledgements.* I wish to thank Dr. Pierre Wijermans, Leyenburg Hospital, The Hague, Netherlands, for helpful discussions and critical reading of the manuscript, and Dr. Roland Mertelsmann for support of clinical and in vitro studies. Supported in part by Wilhelm Sander-Stiftung.

# References

Abele R, Clavel M, Dodion P, Bruntsch U, Gundersen S, Smyth J, Renard J, Van Glabbeke M, Pinedo HM (1987) The EORTC early clinical trials cooperative group experience with azacitidine-2'-deoxycytidine (NSC 127716) in patients with colo-rectal, head and neck, renal carcinomas and malignant melanomas. Eur J Cancer Clin Oncol 23:1921–1924

Aggerholm A, Guldberg P, Hokland M, Hokland P (1999) Extensive intra- and interindividual heterogeneity of p15/INK4B methylation in acute myeloid leukemia. Cancer Res 59:436–441

Andrews F, Nemunaitis J, Tompkins CH, Singer JW (1989) Effect of 5-Azacytidine on gene expression in marrow stromal cells. Mol Cell Biol 9:2748–2751

Anzai H, Frost P, Abruzzese JL (1992) Synergistic cytotoxicity with 2'-deoxy-5-azacytidine and topotecan in vitro and in vivo. Cancer Res 52:2180–2185

Aul C, Gattermann N, Schneider W (1992) Age–related incidence and other epidemiological aspects of myelodysplastic syndromes. Br J Haematol 82:358–367

Avramis VI, Mecum RA, Nyce J, Steele DA, Holcenberg JS (1989) Pharmacodynamic and DNA methylation studies of high-dose-1-β-D-arabinofuranosyl cytosine before and after in vivo 5-azacytidine treatment in pediatric patients with refractory acute lymphocytic leukemia. Cancer Chemother Pharmacol 24:203–210

Baehner RL, Bernstein ID, Sather H, Higgins G, McCreadie S, Chard RL, Hammond D (1979) Improved remission induction rate with D-ZAPO but unimproved remission duration with addition of immunotherapy to chemotherapy in previously untreated children with ANLL. Med Pediatric Oncol 7:127–139

Baylin SB, Herman JG, Vertino PM, Issa JP (1998) Alterations in DNA methylation: a fundamental aspect of neoplasia. Cancer Res 72:141–196

Bender CM, Pao MM, Jones PA (1998) Inhibition of DNA methylation by 5-aza-2'-deoxycytidine suppresses the growth of human tumor cell lines. Cancer Res 58:95–191

Bouchard J, Momparler RL (1983) Incorporation of 5-aza-2'-deoxycytidine 5'-triphosphate into DNA. Interactions with mammalian DNA polymerase and DNA methylase. Mol Pharmacol 24:109–114

Carr BI, Rahbar S, Asmeron Y, Riggs A, Winberg CD (1988) Carcinogenicity and haemoglobin synthesis induction by cytidine analogues. Br J Cancer 57:395–402

Charache S (1997) Mechanism of action of hydroxyurea in the management of sickle cell anemia in adults. Sem Hematol 43(Suppl. 3):15–21

Charache S, Dover G, Smith K, Talbot CC, Moyer M, Boyer S. Treatment of sickle cell anemia with 5-azacytidine results in increased fetal hemoglobin production and is associated with nonrandom hypomethylation of DNA around the γ-δ-β gene complex. Proc Natl Acad Sci USA 80:4842–4846

Chitambar CR, Libnoch JA, Matthaeus WG, Ash RC, Ritch PS. Anderson T (1991) Evaluation of continuous infusion low-dose 5-azacytidine in the treatment of myelodysplastic syndromes. Am J Hematol 37:100–104

Christman JK, Mendelsohn N, Herzog D, Schneiderman N (1983) Effect of 5-azacytidine on differentiation and DNA methylation in human promyelocytic leukemia cells (HL–60). Cancer Res 43: 763–769

Côté S, Momparler RL (1997) Activation of the retinoid acid receptor β gene by 5-aza-2'-deoxycytidine in human DLD-1 colon carcinoma cells. Anti-Cancer Drugs 8:56–61

Cremisi S (1989) Effect of 5-azacytidine treatment on mouse embryonal carcinoma cells. J Cell Physiol 116:181–190

Creusot F, Acs G, Christman JK (1982) Inhibition of DNA methyltransferase and induction of Friend erythroleukemia cell differentation by 5-azacytidine and 5-aza-2'-deoxycytidine. J Biol Chem 2041–2048

Daskalakis M, Nguyen TT, Guldberg P, Wijermans P, Jones PA, Lübbert M (1998) Frequent hypermethylation of P15/INK4B in patients with myelodysplastic syndromes is decreased following treatment with 5-aza-2'-deoxycytidine (decitabine). Blood 92(Suppl. 1):715a

Debusscher L, Marie JP, Dodion P, Blanc GM, Arrigo C, Zittoun R, Stryckmans P (1990) Phase-I–II trial of 5-aza-2'-deoxycytidine in adult patients with acute leukemia. In: 5-Aza-2'-deoxycytidine: preclinical and clinical studies. Eds: RL Momparler and D de Vos. PCH, Haarlem, The Netherlands, p 131–142

D'Incalci M, Covey JM, Zaharko DS, Kohn KW (1985) DNA alkali-labile sites induced by incorporation of 5-aza-2'-deoxycytidine into DNA of mouse leukemia L1210 cells. Cancer Res 45:3197–3202

Dodion PF, Clavel M, Ten Bokkel Huinink W, Robinson E, Renard JF (1990) Phase-II trials with 2'-deoxy-5-azacytidine conducted by the Early Clinical Trials group of the EORTC. In: 5-Aza-2'-deoxycytidine: preclinical and clinical studies. Eds: RL Momparler, D de Vos. PCH, Haarlem, p 117–124

Dover GJ, Charache S, Boyer SH, Moyer M (1985) 5-azacytidine increases HbF production and reduces anemia in sickle cell disease: dose-response analysis of subcutanous and oral dosage regimens. Blood 66:527–532

Dunbar C, Travis W, Kan YW, Nienhuis A (1989) 5-Azacytidine treatment in a β⁰-thalassaemic patient unable to be transfused due to multiple alloantibodies. Br J Haematol 72:467–474

Giralt S, Davis M, O'Brien S, Van Besien K, Champlin R, De Vos D, Kantarjian H (1997) Studies of decitabine with allogeneic progenitor cell transplantation. Leukemia 11 (Suppl 1):32–34

Giralt S, Cohen A, Davis M, O'Brien S, Andersson B, Gajewski J, Khouri I, Körbling M, Champlin R, De Vos D, Kantarjian H (1998a) Phase-I/II study of decitabine with allogeneic peripheral blood stem cell transplantation for treatment of relapse after allogeneic progenitor cell transplantation. Symposium: Treatment of hematologic malignancies with DNA-hypomethylating agents. Miami, FL, December 1998 (abstract)

Giralt S, Cohen A, Davis M, O'Brien S, Andersson B, Gajewski J, Khouri I, Körbling M, Champlin R, De Vos D, Kantarjian H (1998b) Phase-I/II trial combining decitabine with busulfan/cyclophosphamide as a conditioning regimen for allogeneic progenitor cell transplantation. Symposium: Treatment of hematologic malignancies with DNA-hypomethylating agents. Miami, December 1998 (abstract)

Glover AB, Leyland-Jones BR, Chun HG, Davis B, Hoth DF (1987) Azacitidine: 10 years later. Cancer Treatment Rep 71:737–746

Goldberg J, Gryn J, Raza A, Bennett J, Browman G, Bryant J, Grunwald H, Larson R, Vogler R, Preisler H (1993) Mitoxantrone and 5-azacytidine for refractory/relapsed ANLL or CML in blast crisis: a leukemia intergroup study. Am J Hematol 43:286–290

Greenberg P, Cox C, LeBeau MM, Fenaux P, Morel P, Sanz G, Sanz M, Vallespi T, Hamblin T, Oscier D, Ohyashiki K, Toyama K, Aul C, Mufti G, Bennett J (1997) International scoring system for evaluating prognosis in myelodysplastic syndromes. Blood 89:2079–2088

Groeningen CJ, Leyva A, O'Brien Ann MP, Gall HE, Pinedo HM (1986) Phase-I and pharmacokinetic study of 5-aza-2'-deoxycytidine (NSC 127716) in cancer patients. Cancer Res 46:4831–4836

Hellström-Lindberg E, Robert KH, Gahrton G, Lindberg G, Forsblom AM, Kock Y, Ost A (1994) Low-dose ara-C in myelodysplastic syndromes (MDS) and acute leukemia following MDS: proposal for a predictive model. Leuk Lymph 12:343–351

Herman JG, Merlo A, Mao L, Lapidus RG, Issa J-PJ, Davidson NE, Sidransky D, Baylin SB (1995) Inactivation of the CDKN2/p16/MTS1 gene is frequently associated with aberrant DNA methylation in all common human cancers. Cancer Res 55:4525–4530

Herman JG, Civin CI, Issa JP, Collector MI, Sharkis SJ, Baylin SB (1997) Distinct patterns of inactivation of p15INK4B and p16INK4A characterize the major types of hematological malignancies. Cancer Res 57:837–841

Hrodek O, Vesely J (1971) 5-Azacytidine in childhood leukemia. Neoplasma 18:493–503

Humphries RK, Dover G, Young NS, Moore JG, Charache S (1985) 5-Azacytidine acts directly on both erythroid precursors and progenitors to increase production of fetal hemoglobin. J Clin Invest 75:547–557

Jehn U (1994) Long-term outcome of post-remission chemotherapy for adults with acute myeloid leukemia using different dose-intensities. Leuk Lymph 15:99–112

Jehn U, Zittoun R, Suciu S, Fiere D, Haanen C, Peetermans M, Löwenberg B, Willemze R, Solbu G, Stryckmans P, and the EORTC Leukemia Cooperative Group (1990) A randomized comparison of intensive maintenance, treatment for adult acute myelogenous leukemia using either cyclic alternating drugs or repeated courses of the induction-type chemotherapy: AML-6 trial of the EORTC leukemia cooperative group. Haemat Blood Transf 33:277–284

Jones PA, Taylor SM (1980) Cellular differentiation, cytidine analogs and DNA methylation. Cell 20: 85–93

Jones PA, Taylor SM (1981) Hemimethylated duplex DNAs prepared from 5-azacytidine treated cells. Nucleic Acids Res 9:2933–2947

Kantarjian HM, Deisseroth A, Kurzrock R, Estrov Z, Talpaz M (1993) Chronic myelogenous leukemia: a concise update. Blood 82:691–703

Kantarjian HM, O'Brien SM, Estey E, Giralt S, Beran M, Rios MB, Keating M, De Vos D, Talpaz M (1997a) Decitabine studies in chronic and acute myelogenous leukemia. Leukemia 11 (Suppl 1):35–36

Kantarjian HM, O'Brien SM, Keating M, Beran M, Estey E, Giralt S, Kornblau S, Rios MB, De Vos D, Talpaz M (1997b) Results of decitabine therapy in the accelerated and blastic phases of chronic myelogenous leukemia. Leukemia 11:1617–1620

Koeffler HP (1996) Myelodysplastic syndromes. Sem Hematol 33:87–94

Koshy H, Molokie R, Dom L, van der Galien T, Bressler L (1998) Augmentation of fetal hemoglobin (HbF) levels by low dose short duration 5-aza-2'-deoxycytidine (decitabine) administration in sickle cell anemia patients who had no HbF elevation following hydroxyurea therapy. Blood 92 (Suppl 1): 306b

Larson RA, Sweet DL, Harvey M, Golomb HM, Testa JR, Rowley JD (1982) Response to 5-azacytidine in patients with refractory acute non-lymphocytic leukemia and association with chromosome findings. Cancer 49:2222–2225

Lee EJ, Hogge DE, Gallagher R, Schiffer CA (1990) Low-dose 5-azacytidine is ineffective for remission induction in patients with acute myeloid leukemia. Leukemia 4:835–838

Ley TJ, DeSimone J, Anagnou NP, Keller GH, Humphries RK, Turner PH, Young NS, Heller P, Nienhuis AW (1982) 5-Azacytidine selectively increases γ-globin synthesis in a patient with β+ thalassemia. New Engl J Med 307:1469–1475

Ley TJ, DeSimone J, Noguchi CT, Turner PH, Schechter AN, Heller P, Nienhuis AW (1983) 5-Azacytidine increases γ-globin synthesis and reduces the proportion of dense cells in patients with sickle cell anemia. Blood 62:370–380

Li LH, Olin EJ, Buskirk HH, Reineke LM (1974) Cytotoxicity and mode of action of 5-azacytidine on L1210 leukemia. Cancer Res 30:2760–2769

Lowrey CH, Nienhuis AW (1993) Treatment with azacitidine of patients with end-stage β-thalassemia. New Engl J Med 329:845–848

Lübbert M, Haak HL, Kunzmann R, Verhoef G, Wijermans P (1999) Low-dose 5-aza-2'-deoxycytidine (decitabine) in patients with high risk myelodysplastic syndromes: induction of cytogenetic responses. Submitted

Miller KB, Kim K, Morrison FS, Winter JN, Bennett JM, Neiman RS, Head DR, Cassileth PA, O'Connell MJ, Kim K (1992) The evaluation of low-dose cytarabine in the treatment of myelodysplastic syndromes: a phase-III intergroup study. Ann Hematol 65:162–168

Momparler RL (1985) Molecular, cellular and animal pharmacology of 5-aza-2'-deoxycytidine. Pharmacol Ther 30:278–299

Momparler RL, Bouchard J, Onetto N, Rivard GE (1984) 5-Aza-2′-Deoxycytidine therapy in patients with acute leukemia inhibits DNA methylation. Leukemia Res 8:181–185

Momparler RL, Bouchard J, Samson J (1985a) Induction of differentiation and inhibition of DNA methylation in HL-60 myeloid leukemic cells by 5-aza-2′-deoxycytidine. Leuk Res 9:1361–1366

Momparler RL, Rivard GE, Gyger M (1985b) Clinical trial on 5-aza-2′-deoxycytidine in patients with acute leukemia. Pharmacol Therap 30:277–286

Momparler RL, Onetto-Pothier N, Momparler LF (1990) Comparison of antineoplastic activity of cytosine arabinoside and 5-aza-2′-deoxycytidine against human leukemic cells of different phenotype. Leuk Res 14:755–760

Momparler RL, Bouffard DY, Momparler LF, Dionne J, Belanger K, Ayoub J (1997) Pilot phase-I–II study on 5-aza-2′-deoxycytidine (decitabine) in patients with metastatic lung cancer. Anticancer Drugs 8:358–368

Omura GA, Vogler WR, Bartolucci A, Neely CL, Silberman H (1979) Treatment of refractory adult acute leukemia with 5-azacytidine plus β-2′-deoxythioguanosine. Cancer Treat Rep 63:209–210

Petti MC, Mandelli F, Zagonel V, De Gregoris C, Merola MC, Latagliata R, Gattei V, Fazi P, Monfardini S, Pino A (1993) Pilot study of 5-aza-2′-deoxycytidine (decitabine) in the treatment of poor prognosis acute myelogenous leukemia patients: preliminary results. Leukemia 7:36–41

Pinto A, Zagonel V (1993) 5-Aza-2′-deoxycytidine (decitabine) and 5-azacytidine in the treatment of acute myeloid leukemias and myelodysplastic syndromes: past, present and future trends. Leukemia 7(Suppl 1):51–60

Pinto A, Attadia V, Fusco A, Ferrara F, Spada OA, Di Fiore PP (1984a) 5-Aza-2′-deoxycytidine induces terminal differentiation of leukemic blasts from patients with acute myeloid leukemias. Blood 64: 922–929

Pinto A, Maio M, Attadia V, Zappacosta S, Cimino R (1984b) Modulation of HLA-DR antigen expression in human myeloid leukaemia cells by cytarabine and 5-aza-2′-deoxycytidine. Lancet II(8407):867–868

Pinto A, Zagonel V, Attadia V, Bullian PL, Gattei V, Carbone A, Monfardini S, Colombatti A (1989) 5-Aza-2′-deoxycytidine as a differentiation inducer in acute myeloid leukaemias and myelodysplastic syndromes of the elderly. Bone Marrow Transplant Suppl 3:28–32

Pinto A, Zagonel V, Gattei V, Marotta G, Bullian PL, Mancardi S, Coglievina M, De Rosa L, Alosi M, Carbone A, Attadia V (1990) 5-Aza-2′-deoxycytidine as a differentiation inducer in human hemopoietic malignancies: preliminary observations on the in vivo modulation of leukemia cells phenotype and correlations with clinical response. In: 5-Aza-2′-deoxycytidine: preclinical and clinical studies. Eds: RL Momparler, D de Vos. PCH, Haarlem, p 143–164

Piskala A, Sorm F (1964) Nucleic acids components and their analogues. LI. Synthesis of 1-glycosyl derivatives of 5-azauracil and 5-azacytosine. Collect Czeck Chem Commun 29:2060–2076

Pliml J, Sorm F (1964) Synthesis of 2′-deoxy-D-ribofuranosyl-5-azacytosine. Coll Czeck Chem Commun 29:2576–2577

Quesnel B, Guillerm G, Vereecque R, Wattel E, Preudhomme C, Bauters F, Vanrumbeke M, Fenaux P (1998) Methylation of the p15$^{INK4B}$ gene in myelodysplastic syndromes is frequent and acquired during disease progression. Blood 91:2985–2990

Ramchandani S, MacLeod AR, Pinard M, Von Hofe E, Szyf M (1997) Inhibition of tumorigenesis by a cytosine-DNA, methyltransferase, antisense oligonucleotide. Proc Natl Adac Sci USA 94:684–689

Richel DJ, Colly LP, Kluin-Nelemans JC, Willemze R (1991) The antileukemic activity of 5-aza-2′-deoxycytidine (Aza-dC) in patients with relapsed and resistant leukemia. Br J Cancer 64:144–148

Rivard GE, Momparler RL, Demers J, Benoit P, Raymond R, Lin K, Momparler LF (1981) Phase I study on 5-aza-2′-deoxycytidine in children with acute leukemia. Leuk Res 5:453–462

Rugo H, Damon L, Ries C, Linker C (1999) Compassionate use of subcutaneous 5-azacytidine (AzaC) in the treatment of myelodysplastic syndromes (MDS). Leuk Res 23 (Suppl 1):72

Sacchi S, Talpaz M, O'Brien S, Cortes J, Kantarjian H (1998) decitabine, a hypomethylating agent, is active for the treatment of chronic myelogenous leukemia (CML) in non-lymphoid blastic phase (BP). Blood 92 (Suppl 1):252a (abstract)

Saiki JH, Bodey GP, Hewlett JS, Amare M, Morrison FS, Wilson HE, Linman JW (1981) Effect of schedule on activity and toxicity of 5-azacytidine in acute leukemia: a Southwest Oncology Group study. Cancer 47:1739–1742

Saiki JH, McCredie KB, Vietti TJ (1987) 5-azacytidine in acute leukemia. Cancer 42:2111–2114

Santi DV, Garret CD, Barr PJ (1983) On the mechanism of inhibition of DNA–cytosine methyltransferase by cytidine analogs. Cell 83:83–89

Schiffer CA, DeBellis R, Kasdorf H, Wiernik PH (1982) Treatment of the blast crisis of chronic myelogenous leukemia with azacitidine and VP-16-213. Cancer Treat Rep 66:267–271

Schwartsmann G, Fernandes MS, Schaan MD, Moschen M, Gerhardt LM, DiLeone L, Loitzembauer B, Kalakun L (1997) Decitabine (5-aza-2'-deoxycytidine; decitabine) plus daunorubicin as a first line treatment in patients with acute myeloid leukemia: preliminary observations. Leukemia 11 (Suppl 1): 28–31

Sessa C, Bokkel Huinink WT, Stoter G, Renards J, Cavalli F (1990) Phase-II study of 5-aza-2'-deoxycytidine in advanced ovarian carcinoma. Eur J Cancer 26:137–138

Shadduck RK, Lister J, Raymond JM, Zeigler ZR (1999) 5-azacytidine therapy for myelodysplasia. Leuk Res 23 (Suppl 1):72 (abstract)

Silverman LR, Holland JF, Weinberg RS, Aller BP, Davis RB, Ellison RR, Demakos EP, Cornel CJ, Carey R, Schiffer C, Frei E, McIntyre O (1993) Effect of treatment with 5-Azacytidine on the in vitro and in vivo haematopoiesis in patients with myelodysplastic syndromes. Leukemia 7 (Suppl 1): 21–29

Silverman LR, Holland JF, Demakos EP, Perterson P, Nelson DA, Clamon G, Powell BL, Larson R, Bloomfield CD, McIntyre Or, Schiffer C (1994) Azacitidine (AzaC) in myelodysplastic syndromes (MDS), CALGB Studies 8421 and 8921. Ann Hemat 68 (Suppl 2):21a (abstract)

Silverman LR, Demakos EP, Peterson B, Odchimar-Reissig R, Nelson D, Kornblith AB, Stone R, Holland JC, Powell BL, DeCastro C, Ellerton J, Larson RA, Schiffer CA, Holland JF (1998) A randomized controlled trial of subcutaneous azacitidine (AzaC) in patients with the myelodysplastic syndrome (MDS): a study of the Cancer and Leukemia Group B (CALGB). Proc Am Soc Clin Onc 17:14a (abstract)

Sorm F, Vesely J (1964) The activity of a new antimetabolite, 5-azacytidine, against lymphoid leukaemia in AK mice. Neoplasma 11:123–130

Taylor SM, Jones PA (1982) Changes in phenotypic expression in embryonic and adult cells treated with 5-azacytidine. J Cell Physiol 111:187–194

Thibault A, Figg WD, Bergan RC, Lush RM, Myers CE, Tompkins A, Reed E, Samid D (1998) A phase-II study of 5-aza-2'-deoxycytidine (decitabine) in hormone-independent metastatic (D2) prostate cancer. Tumori 84:87–89

Torrealba-de Ron AT, Papayannopoulou T, Knapp MS, Fu MF, Knitter G, Stamatoyannopoulos G (1984) Perturbations in the erythroid marrow progenitor cell pools may play a role in the augmentation of HbF by 5-azacytidine. Blood 63:201–210

Uchida T, Kinoshita T, Nagai H, Nakahara Y, Saito H, Hotta T, Murata T (1997) Hypermethylation of the p15INK4B gene in myelodysplastic syndromes. Blood 90:1403–1409

Van der Ploeg LH, Flavell RA (1980) DNA methylation in the human γ-δ-β-globin locus in erythroid and nonerythroid tissues. Cell 19:947–958

Vermorken JB, Tumolo S, Roozendaal KJ, Guastalla JP, Splinter Ted AW, Renard J (1991) 5-Aza-2'-deoxycytidine in advanced or recurrent cancer of the uterine cervix. Eur J Cancer 27:216–217

Vogler WR, Miller DS, Keller JW (1976) 5-Azacytidine (NSC 102816): a new drug for the treatment of myeloblastic leukemia. Blood 48:331–347

Vogler WR, Winton EF, Gordon DS, Raney MR, Go B, Meyer L (1984) A randomized comparison of postremission therapy in acute myelogenous leukemia: a Southeastern Cancer Study Group trial. Blood 63:1039–1045

Vogler WR, Weiner RS, Moore JO, Omura GA, Bartolucci AA, Stagg M (1995) Long-term follow-up of a randomized post-induction therapy trial in acute myelogenous leukemia (a Southeastern Cancer Study Group trial). Leukemia 9:1456–1460

Von Hoff DD, Slavik M, Muggia FM (1976) A new anticancer drug with effectiveness in acute myelogenous leukemia. Ann Intern Med 85:237–245

Wijermans PW, Krulder JW, Huijgens PC, Neve P (1997) Continuous infusion of low-dose 5-aza-2'-deoxycytidine in elderly patients with high-risk myelodysplastic syndrome. Leukemia 11:1–5

Wijermans P, Lübbert M, Verhoef G, Huijgens P, Ravoet C, Andre M, Ferrant A (1999) Low-dose 5-aza-2'-deoxycytidine (decitabine), a DNA hypomethylation agent, for the treatment of high risk myelodysplastic syndrome; a multicenter phase-II study in elderly patients. J Clin Oncol (in press)

Willemze R, Archimbaud E, Muus P (1993) Preliminary results with 5-aza-2'deoxycytidine (DAC)-containing chemotherapy in patients with relapsed or refractory acute leukemia. The EORTC Leukemia Cooperative Group. Leukemia 7 (suppl 1):49 50

Willemze R, Suciu S, Archimbaud E, Muus P, Stryckmans P, Louwagie EA, Berneman Z, Tjean M, Wijermans P, Döhner H, Jehn U, Labar B, Jaksic B, Dardenne M, Zittoun R (1997) A randomized phase-II study on the effects of 5-aza-2'-deoxycytidine combined with either amsacrine or idarubicin in patients with relapsed acute leukemia: an EORTC Leukemia Cooperative Group phase-II study. Leukemia 11 (Suppl 1):24–27

Wilson Vl, Jones PA, Momparler RL (1983) Inhibition of DNA methylation in L1210 leukemia cells by 5-aza-2'-deoxycytidine as a possible mechanism of chemotherapeutic action. Cancer Res 43:3493–3496

Zagonel V, Pinto A, Bullian PL, Sorio R, Gattei V, Curri G, De Rosa L, Attadia V, Monfardini S (1990) 5-Aza-2'-deoxycytidine as a differentiation inducer in human hemopoietic malignancies: preliminary clinical report. In: 5-Aza-2'-deoxycytidine: preclinical and clinical studies. Editors: Richard L. Momparler and Dick de Vos. PCH, Haarlem, Netherlands, 165–181

Zagonel V, Lo Re G, Marotta G, Barbare R, Sardeo R, Gattei V, De Angelis V, Montardinni S, Pinto A (1993) 5-Aza-2'-deoxycytidine (decitabine) induces trilineage response in unfavourable myelodysplastic syndrome. Leukemia (Suppl 1):30–35

Zingg JM, Jones PA (1997) Genetic and epigenetic aspects of DNA methylation on genome expression, evolution, mutation and carcinogenesis. Carcinogenesis 18:869–882

# Subject Index

# Current Topics in Microbiology and Immunology

Volumes published since 1989 (and still available)

Vol. 226: **Koprowski, Hilary; Weiner, David B. (Eds.):** DNA Vaccination/Genetic Vaccination. 1998. 31 figs. XVIII, 198 pp. ISBN 3-540-63392-8

Vol. 227: **Vogt, Peter K.; Reed, Steven I. (Eds.):** Cyclin Dependent Kinase (CDK) Inhibitors. 1998. 15 figs. XII. 169 pp. ISBN 3-540-63429-0

Vol. 228: **Pawson, Anthony I. (Ed.):** Protein Modules in Signal Transduction. 1998. 42 figs. IX, 368 pp. ISBN 3-540-63396-0

Vol. 229: **Kelsoe, Garnett; Flajnik, Martin (Eds.):** Somatic Diversification of Immune Responses. 1998. 38 figs. IX, 221 pp. ISBN 3-540-63608-0

Vol. 230: **Kärre, Klas; Colonna, Marco (Eds.):** Specificity, Function. and Development of NK Cells. 1998. 22 figs. IX, 248 pp. ISBN 3-540-63941-1

Vol. 231: **Holzmann, Bernhard; Wagner, Hermann (Eds.):** Leukocyte Integrins in the Immune System and Malignant Disease. 1998. 40 figs. XIII, 189 pp. ISBN 3-540-63609-9

Vol. 232: **Whitton, J. Lindsay (Ed.):** Antigen Presentation. 1998. 11 figs. IX, 244 pp. ISBN 3-540-63813-X

Vol. 233/I: **Tyler, Kenneth L.; Oldstone, Michael B. A. (Eds.):** Reoviruses I. 1998. 29 figs. XVIII, 223 pp. ISBN 3-540-63946-2

Vol. 233/II: **Tyler, Kenneth L.; Oldstone, Michael B. A. (Eds.):** Reoviruses II. 1998. 45 figs. XVI, 187 pp. ISBN 3-540-63947-0

Vol. 234: **Frankel, Arthur E. (Ed.):** Clinical Applications of Immunotoxins. 1999. 16 figs. IX, 122 pp. ISBN 3-540-64097-5

Vol. 235: **Klenk, Hans-Dieter (Ed.):** Marburg and Ebola Viruses. 1999. 34 figs. XI, 225 pp. ISBN 3-540-64729-5

Vol. 236. **Kraehenbuhl, Jean-Pierre; Neutra, Marian R. (Eds.):** Defense of Mucosal Surfaces: Pathogenesis, Immunity and Vaccines. 1999. 30 figs. IX, 296 pp. ISBN 3-540-64730-9

Vol. 237: **Claesson-Welsh, Lena (Ed.):** Vascular Growth Factors and Angiogenesis. 1999. 36 figs. X, 189 pp. ISBN 3-540-64731-7

Vol. 238: **Coffman, Robert L.; Romagnani, Sergio (Eds.):** Redirection of Th1 and Th2 Responses. 1999. 6 figs. IX, 148 pp. ISBN 3-540-65048-2

Vol. 239: **Vogt, Peter K.; Jackson, Andrew O. (Eds.):** Satellites and Defective Viral RNAs. 1999. 39 figs. XVI, 179 pp. ISBN 3-540-65049-0

Vol. 240: **Hammond, John; McGarvey, Peter; Yusibov, Vidadi (Eds.):** Plant Biotechnology. 1999. 12 figs. XII, 196 pp. ISBN 3-540-65104-7

Vol. 241: **Westblom, Tore U.; Czinn, Steven J.; Nedrud, John G. (Eds.):** Gastroduodenal Disease and Helicobacter pylori. 1999. 35 figs. XI, 313 pp. ISBN 3-540-65084-9

Vol. 242: **Hagedorn, Curt H.; Rice, Charles M. (Eds.):** The Hepatitis C Viruses. 2000. 47 figs. IX, 379 pp. ISBN 3-540-65358-9

Vol. 243: **Famulok, Michael; Winnacker, Ernst-L.; Wong, Chi-Huey (Eds.):** Combinatorial Chemistry in Biology. 1999. 48 figs. IX, 189 pp. ISBN 3-540-65704-5

Vol. 244: **Daëron, Marc; Vivier, Eric (Eds.):** Immunoreceptor Tyrosine-Based Inhibition Motifs. 1999. 20 figs. VIII, 179 pp. ISBN 3-540-65789-4

Vol. 245/I: **Justement, Louis B.; Siminovitch, Katherine A. (Eds.):** Signal Transduction and the Coordination of B Lymphocyte Development and Function I. 2000. 22 figs. XVI, 274 pp. ISBN 3-540-66002-X

Vol. 245/II: **Justement, Louis B.; Siminovitch, Katherine A. (Eds.):** Signal Transduction on the Coordination of B Lymphocyte Development and Function II. 2000. 13 figs. XV, 172 pp. ISBN 3-540-66003-8

Vol. 246: **Melchers, Fritz; Potter, Michael (Eds.):** Mechanisms of B Cell Neoplasia 1998. 1999. 111 figs. XXIX, 415 pp. ISBN 3-540-65759-2